ACADEMIA FOR GREEN AFRICA

www.afgainitiatives.org

Building Environmental Competence in Africa

AFRICAN ENVIRONMENTAL PERSPECTIVES

Volume 1

AFRICAN ENVIRONMENTAL PERSPECTIVES

VOLUME 1

AN ACADEMIA FOR GREEN AFRICA
PUBLICATION

Dr. Akanimo Odon and Dr. Sam Guobadia

authorHOUSE®

AuthorHouse™
1663 Liberty Drive
Bloomington, IN 47403
www.authorhouse.com
Phone: 1-800-839-8640

First published by AuthorHouse   06/16/2011

ISBN: 978-1-4567-8444-7 (sc)
ISBN: 978-1-4567-8445-4 (ebk)

Printed in the United States of America

Any people depicted in stock imagery provided by Thinkstock are models, and such images are being used for illustrative purposes only.
Certain stock imagery © Thinkstock.

This book is printed on acid-free paper.

For contact and more information on Academia for Green Africa (AFGA)
Initiatives and on how to get involved, contact the founders on:

Email: info@envirofly-group.com

Check the AFGA Website for details of programmes, projects and events

Website: www.afgainitiatives.org

TOWARDS A GREENER AFRICA:

The Role of Academic Institutions in Capacity Building

October 19th - 21st, 2010

Benin City, Nigeria

Contents

The issue of the environment is undoubtedly the most pressing issue facing mankind. Although many industrialized countries have encouraged widespread commentary on the subject, Africa and other developing countries seem to be pre-occupied with the issues of economic sustainability and, as a result, these countries have devoted less time to the very important issues of climate change, environmental degradation, waste management, etc. For this reason, Envirofly Group and Benson Idahosa University raised the level of discourse on the subject matter at the International Conference that held in Africa in **October of the year 2010**.

Given the devastating impacts of environmental degradation in African countries as well as other parts of the world, the International Conference is aimed at rigorously addressing the economic, political, social, legal, historical and the scientific questions which should focus on an appropriate response to the threats of global warming and climate change, particularly, with respect to economic and social development in Sub-Saharan Africa. This has become imperative in view of the lackluster response to these problems in Africa.

The discussion among most enlightened people everywhere today centers on climate change and how it portends massive alterations in weather patterns all over the world with the consequent social and economic disruptions. The questions are: What are the appropriate roles of academic institutions and the African state in environmental management in a rapidly degrading environment? How will Africa manage the consequences of environmental fall outs? It is desirable for African institutions and governments to intervene in building awareness of the effects of environmental degradation in order to provide institutional arrangements and safeguards necessary to ensure better waste management, use of renewable energy and greater public awareness.

The Academia for Green Africa (AFGA) is an initiative of Envirofly Group, United Kingdom and Benson Idahosa University, Benin City, Nigeria that identifies the role of academic institutions in Africa towards meeting and addressing the current global challenges of climate Change and the Environment. Conceived in early 2010 as a follow up to the deliberations of the last Climate Change Summit in Copenhagen, this initiative is strategic in bridging the environmental competence gaps currently in Africa using strategic partnerships and networks with established environmental academic and corporate institutions and bodies globally.

AFGA plans to host regular academic conferences and symposia with a view of providing for intense discussions of environmental issues and constantly encourage *Green Initiatives* under the supervision of International Advisory and Governing Boards. These academic exercises will be complemented by a host of press bulletins and academic publications, notably, the *African Environmental Perspectives*, which will be a leading reference material on African environmental issues, concerns and challenges.

Dr. Sam Guobadia   Dr, Akanimo Odon
Co-Founder, AFGA   Co-Founder, AFGA
&   &
Chairman, LOC   CEO, Envirofly Group
AFGA 2010, Benson Idahosa Unversity

2.0 WELCOME ADDRESS BY THE VICE CHANCELLOR, PROFESSOR GIDEON E. D. OMUTA, AT THE OPENING CEREMONY OF THE FIRST INTERNATIONAL CONFERENCE OF ACADEMIA FOR GREEN AFRICA (AFGA), ON TUESDAY, 19TH OCTOBER, 2010

Your Excellency, Comrade Adams Aliyu Oshiomhole, Executive Governor of Edo State of Nigeria;

Honourable Mohammed Ka'oje Abubakar, Honourable Minister of Science & Technology;

Honourable John Ogar Odey, Honourable Minister of Environment;

Honourable Godsdey Peter Orubebe, Honourable Minister of Niger Delta Affairs;

Professor Julius Okojie, Executive Secretary of the National Universities Commission;

Our Distinguished Overseas Guests;

Professors and other distinguished paper presenters here present;

Other distinguished invitees;

Students and researchers;

Gentlemen of the Press;

Ladies and Gentlemen.

It is a very unique honour and privilege for me, as Vice-Chancellor of Benson Idahosa University, to welcome all of you to this first International Conference of **Academia for Green Africa (AFGA),** which is taking place here in Benin City from today, Tuesday, 19th to Thursday, 21st October, 2010. AFGA is an initiative of Envirofly Group in collaboration with Benson Idahosa University. Benson Idahosa University is a young private university which is driven by the desire and passion to provide outstanding academic, professional and entrepreneurial training imbued with strong moral and spiritual code, such as will enable its products to instigate change, foster progress and instil spiritual growth in the larger society. The university's core purpose is: '**Change Nigeria**'. It seeks to accomplish this through not only its well entrenched and robust teaching and research programmes but also through a number of community-driven programmes; the AFGA initiative being one of them.

AFGA aims at providing fora for harnessing the efforts of the academia (local, national and international), towards developing appropriate responses to the looming dangers posed to mankind in general, by the gradual and persistent assault on, and consequent deterioration of, the environment, all over the world, but with special focus on Africa. Of the numerous threats which confront humanity today, the ones to the environment are perhaps the most urgent and compelling, which are manifested in various ways, including; air pollution, water pollution, land pollution, despoliation or degradation, soil erosion, deforestation and desertification, destruction of animal resources, biodiversity loss, flooding and numerous other deleterious actions of man, and their inestimable collateral damages. Many of these occurrences result in green house emissions which, in turn, cause now well-known global warming and climate change. The torrential rains resulting in devastating floods and landslides, crippling droughts, destructive hurricanes and typhoons, which have been occurring in various parts of planet earth; with increasing frequency of late and which cause devastation, death and displacement of populations, are all eloquent testimonies of climate change. These incidents certainly compel all of us to feel concerned and challenged.

AFGA desires to be an active player in the general efforts to develop appropriate response strategies, capacities and competences for meeting the challenges posed by the various parameters of climate change. It recognises the fact that the environment (and in particular climate change) presents varied and complex dysfunctions which draw from a variety of intellectual disciplines to find appropriate solutions and mitigations. It draws attention to the grim possibility that planet earth may one day cease to provide conducive conditions to sustain human (nay, all organic) life, if the present trends do not abate or are not moderated and checked. It also recognises the fact that the issues involved raise questions to which there are no easy answers. This is clearly confirmed by the fact that numerous summits have taken place in the past (Rio de Janeiro, 1992; Kyoto, 1997, and recently, Copenhagen, 2009) which have sought to agree on a common strategy for stemming the threats of climate change, without substantial success.

It is curious that, while the level of societal awareness, sensitivity and concern has been raised to quite appreciable heights in many parts of the developed world, this is not yet so in the less developed countries, especially, those of Africa. In Africa, the people are so overwhelmed with the struggle for survival that they do not seem to be aware that a problem exists and consequently do not seem to be conscious of, and concerned with, the numerous ways in which their actions or inactions have contributed, and are contributing, to global warming, nor the effects which global warming may be having on the quality of their lives.

An example of this wide disparity in the level of consciousness and awareness can be seen from hue and cry which greeted the recent Deepwater Horizon oil rig accident which occurred in the Gulf of Mexico, leading to the spill of four million of crude. You will recall that the attention of the whole world was aroused by this incident and British Petroleum was made to bear the enormous cost of remediating the neighbouring sea shores and to compensate all those whose means of livelihood were affected. It is on record that this cost the company well over $26 billion.

Now the questions which this scenario raises are: What about the enormous devastation and destitution from the over 68,000 reported oil spills involving over 13 million barrels, which have been going on in the Niger Delta in Nigeria, since oil was first drilled in commercial quantity on June 5, 1956 in Oloibiri, and the first barrels of oil were exported from that well in 1958? How much have these incidents cost the multinational oil companies in compensation?

It bothers one why the age long Niger Delta disasters have not raised anything compared by the level of concern raised in response to the Gulf of Mexico oil spill! For instance, on September 29, 2010, BP and the Gulf of Mexico Alliance announced a $500 million (or ₦75 billion) independent research initiative to study the effects of the Deepwater Horizon incident and potential associated impact on the environment and public health. Again, the question is: How much have SPDC, Chevron, Total, Agip, Mobil and the others committed to the study of the impact of the spills that their operations have caused over the years, in the Niger Delta region?

The difference in the nature of response generated by the single Gulf of Mexico incident, on the one hand, and the scores of thousands of the incidents in Nigeria, on the other, is clearly attributable to the level of awareness.

Everything, therefore, points to the need to begin to stimulate interest in the subject of the environment in this part of the world. AFGA seeks to fill this niche by involving all concerned, not just the academia, in the efforts to create the necessary awareness. This is the principal aim of this first conference. But it will certainly not end there. It is anticipated that the conference will be followed later by concrete measures to address the challenges, through research studies and publications, which will show not only the nature of our environmental challenges, and the manner in which they manifest, but also point to the various ways in which these challenges can be mitigated.

Through AFGA, we in Benson Idahosa University aspire to lead in this endeavour. AFGA recognises that the problem is multifaceted, and that it correspondingly involves many disciplines of study. It also recognises the fact that appropriate solutions will involve all levels of society, including in particular, government, which must be assisted to formulate the right policies, to establish appropriate supervisory agencies and to provide the required funding. Being essentially an awareness creating endeavour, this first conference has been designed to cover a wide spectrum of environmental issues, ranging from waste management to renewable sources of energy, from environmental economics and finance to environmental education, from environmental law and ethics to gender issues, to name just a few. Consequently, different stakeholders have been invited to participate; policy makers (i.e., government: federal, state and local), the academia (all universities in the land and from overseas), the National Universities Commission, etc., the media, and the general public. We are fortunate to have gotten the cooperation of some foreign experts (from the UK) who will be presenting papers. Our effort to bring experts from India were frustrated by minor last-minute challenges in the Nigerian embassy.

To all of you who have come to share knowledge on this all important subject, and all others who have responded to our invitation by their presence here this morning, we say, a very warm welcome. It is our hope and expectation that a solid foundation will be laid at this conference from which we can launch further initiatives aimed at addressing the more specialized and topical challenges of global warming and climate change.

On behalf of Benson Idahosa University and our partner, Envirofly Group, and in particular, Academia for Green Africa (AFGA), I welcome you all to Benin City and I wish the conference very successful deliberations.

God bless you.

3.0 A KEYNOTE ADDRESS DELIVERED BY THE HONOURABLE MINISTER OF SCIENCE AND TECHNOLOGY, PROF. MOHAMMED KA'OJE ABUBAKAR AT THE 1ST ACADEMIA FOR GREEN AFRICA (AGFA) INTERNATIONAL CONFERENCE HELD AT THE BENSON IDAHOSA UNIVERSITY, BENIN CITY.

Ladies and Gentlemen, I am indeed delighted and at the same time grateful that the **Academia for Green Africa (AFGA)** deemed it fit to invite me to present a Keynote Address at this International Conference. I wish to thank in particular the Benson Idahosa University and Envirofly Consulting UK Ltd for using this auspicious gathering to raise the level of discourse on this subject matter.

2. Given the devastating impacts of environmental degradation on African countries, as well as other parts of the world, the International Conference I am told, is aimed at rigourously addressing the economic, political, social, legal, historical and the scientific questions which focus on an appropriate response to the threats of global warming and climate change, particularly with respect to economic and social development in Africa.

3. I am aware that several preparatory meetings are going on in the entire world to suggest and map a way forward towards combating the fall-outs of climate change. In these processes, the academic institutions: Universities, Research Institutes, Polytechnics and other elite organizations in both the public and private sectors are expected to work in concert to find appropriate solutions through technological tools to enable Africa address the adverse effects of Climate Change.

4. If nations like the USA which initially did not accede to the Kyoto Protocol are now coming on board and admitting to the reality as well as assuming frontline positions in the current negotiations on Climate Change, Africa which contributes to climate change through oil exploration/exploitation and the attendant gas flaring from such activities, as well as impact from the extractive and manufacturing industries, cannot afford to remain in the background particularly, with the on-going efforts to reach an agreement that will outlive the Kyoto Protocol come 2012 which will be binding on all nations for the next 50 years.

5. According to the United States President, Mr. Barrack Obama, ". . . . *All across the world, in every kind of environment and region known to man, increasingly dangerous weather pattern and devastating storms are abruptly putting an end to the long-running debate over whether or not climate change is real. Not only is it real, its here, and its effects are giving rise to frightening new global phenomenon: the man-made natural disaster*".

6. As it were, the parlance in Africa had been rustic. There is indeed a need to re-green it.

7. At this juncture, it is pertinent to examine the current environmental situation in the Africa Continent.

8. As pointed out in the preamble to the Lagos Plan of Action, Africa is unable to achieve any significant growth rate or satisfactory index of general well-being in the past 40 years. Whatever socio-economic indicator is used—be it per capital income, the share of primary activities in total production, school enrolment ratios, access to safe water, mortality or health or the entire gamut contained in the New Partnership for Africa Development (NEPAD), or for us in Nigeria, the Millennium Development Goals (MDGs) or the Vision 20:2020—most African countries can be seen to be lagging behind other developing countries. The number of African countries listed as "least developed" by the United Nations recently increased to 26 out of a world total of 36, while 20 out of 33 countries classified by the World Bank as "low income" developing countries are located in Africa. The share of manufacturing in the region's GDP is still appreciably lower than the comparable average for all other developing countries, while agricultural performance has dropped drastically, bearing little comparison with the previous decades or with performance in other developing regions. Given the close link between agriculture and industry, poor performance in the agricultural sector has devolved negatively upon manufacturing.

9. These economic difficulties are compounded by the persistent balance of payment deficits faced by most countries in the region: the external debt of the region increased five-fold during the past four decades while external reserves dropped to critically low levels. The expansion of manufacturing output in the region is also hampered by sluggish domestic markets, inadequate raw material supplies for key industries, the absence of skilled and experienced industrial manpower, and shortages of imported materials, spare parts and machinery. The situation is further aggravated by major difficulties stemming from the energy problems facing the region despite its substantial energy potential. The inadequacy of the region's transport and communications infrastructure coupled with the inefficiency of the services sectors are also recognized as major obstacles to the socio-economic development of the region.

10. The generally stagnant nature of the domestic economies has inevitably depressed industrial investment and, in turn, future expansion. The fact that the typical African economy is still at an early stage of development means that certain "structural" features condition the environment in which industry operates. On the positive angle, there is a potential for industrial development particularly with the abundant rich natural resource endowment of many African countries. On the negative side, however, the small populations and low levels of income in most African countries mean that existing domestic markets for consumer goods are limited and in fact, far too small to permit the attainment of maximum economies of scale in many branches of industry.

11. The effects of unrealised global development strategies which are more sharply felt in Africa than in other continents of the world have compelled Africa to undertake basic restructuring of its economic base. It will be recalled that the main objective of economic development in most African countries since independence, has been to achieve a sustained increase in the standard of living for its increasing population. In order to achieve this, the emphasis must shift from primary production to secondary activities, i.e to industrialize. To this long-term strategy of industrialization can be added the shorter term goals of an accelerated growth in output and the creation of employment opportunities so as to reduce unemployment or under-employment and contribute to the elimination of mass poverty. Thus far, I have enumerated the structural constraints to Africa's development. On the other hand, Climate Change is becoming another

critical developmental issue and I hope Africans will see the opportunities inherent in its challenges.

12. Climate change is becoming a potential threat not only to the socio-economic activities of nations but more importantly to human existence. It is being felt across the globe, and is mainly caused by human activities which have led to atmospheric changes. It is expected to continue for decades irrespective of how rigorous mitigation efforts might be. This constitutes a serious challenge to our academics who are expected to proffer appropriate solutions.

13. Though Climate Change affects everyone, it is expected to have a disproportionate effect on the poor living in developing countries of the world. Some of these include Africa, Asia, Latin America and the Small Island States. This is because many of these areas are in the low lying regions of the world and therefore, subject to coastal flooding. In addition, a large percentage of the population of these areas survive on agriculture, which is increasingly threatened by global warming and therefore, indirectly, constituting a threat to food security.

14. The effects of climate change on people, particularly the poor, can be severe. Farmers, pastoralists, fishing communities and town dwellers are vulnerable to changes in water availability and lower agricultural productivity. Warmer climate increases the risk of contracting vector-borne diseases such as malaria. The economic implications of these changes are enormous. Receipts from agricultural activities which account for over half the jobs and GDP in Africa may decline sharply. Thus, as national revenues are strained, demand for public expenditures will increase. Climate Change also has the potential of undermining the achievements of the Millennium Development Goals such as the eradication of extreme poverty and hunger by 2015.

15. There is a deep concern arising from the growing body of scientific evidences as mentioned in projections and predictions from the Inter-governmental Panel on Climate Change (IPCC) in its 4th Assessment Report (AR4), which clearly states that Africa is the most vulnerable to the adverse impacts of Climate Change. This is grossly disproportionate as Africa contributes less than 3% to global emissions, which is more so because it lacks the capacity to cope with natural disasters and will continue to pay heavy toll in causalities and destroyed infrastructure.

16. Ladies and Gentlemen, let me challenge you to focus in your deliberations, the following important issues. These are:

- "Greening";
- Green Technology which provides new opportunities for Greening Africa;
- Conventional Remedies Systemwide and Ecological Initiatives;
- Green Economy;
- Biodiversity;
- Ecosystem;
- Negative Externalities which explain the activities that impose cost on the masses while greening the economy and
- Consequences of Inaction.

17. As you might be probably aware, my Ministry shares the responsibility for climate change and has accordingly initiated some actions towards the amelioration of its adverse effects. These include:

- **Clean Development Mechanism (CDM)**—The purpose of the CDM is to assist Parties not included in Annex I countries in achieving sustainable development and in contributing to the ultimate objective of the Kyoto Protocol in the Convention on Climate Change and to assist Parties included in Annex I in achieving compliance with their quantified emission limitation and reduction commitments under Article 5. The Ministry has been an active participant in the Carbon Trade Negotiations.

- **Bioremediation**—In this regard, the National Biotechnology Development Agency (NABDA) which is an Agency under my purview has been active in developing various bioengineered organisms. For example, bioengineered bacteria are being employed to extract copper from galvanizing-caused wastes and to halve the amount of hydrogen peroxide used in bleaching paper in order to slash the cost of manufacturing newsprints, coffee filters and to detect and decompose TNT and other contaminants of our soil and water.

- **Tissue Culture**—It has also been involved in techniques aimed at massively producing plantlets of elite food and horticultural crops and plants. These plantlets are subsequently used in Afforestation thereby, enhancing carbon sequestration.

- **Satellite Data**—As you are aware, the National Space Research and Development Agency (NASRDA) has taken giant strides in Satellite data applications in:

  (a) Food Security:-

  - In the development of Early Warning Systems for Food Security in Nigeria using NigeriaSat-1 and other Satellites and also,
  - In the development of Fadama Land Information Management System (FLIMS) in collaboration with Space Application Centre South Africa.

  (b) Resources Inventory Management:

  - Mapping of Settlements and Major Roads in Nigeria.
  - Using Remote Sensing (RS), Geo-Information Systems (GIS) and Geo-Positioning System (GPS) technologies for the identification of Artisanal and Illegal mining sites in Nigeria.

  (c) Ecology and Disaster Management:

  - Satellite-based Environmental Change Research in the Niger Delta in collaboration with the Laboratory for Climate Change of Missouri, Kansas city (UMKC), USA. Phase 1 of the project has been completed. The Niger Delta Development Commission (NDDC) and Shell Petroleum have indicated their interest in participating in the second phase of the project.

- Deforestation in Nigeria with Implications on Biodiversity. A Geoinformation System-Based Forest Monitoring in Nigeria in collaboration with OAU, Ile-Ife and International Institute for Geo-Information Science and Earth Observation (ITC), The Netherlands.
- A spatio-temporal Assessment of Climate and Human Induced Impact on Ecosystems Degradation and Water Resources management around Kainji Lake area in collaboration with FUT Minna.
- Development of RS and GIS Predictive Models for Desertification Early Warning in collaboration with Federal University of Technology Yola and North East Arid Zone Project and Federal Ministry of Environment.
- Mapping and Monitoring of the Impact of Gully Erosion in South Eastern Nigeria in collaboration with Nnamdi Azikwe University, Awka.
- Flood Disaster Vulnerability Mapping and Monitoring in Nigeria in Collaboration with National Emergency Management Agency (NEMA).

18. I wish to conclude by alerting this august gathering about the dangers inherent in Climate Change and the urgent need for the Academia to map out relevant strategies to combat the menace. To do otherwise, will be tantamount to complacency. This forum therefore, presents us with a unique opportunity to fully deliberate on this very important environmental issue.

I wish you very fruitful deliberations.

4.0 NEED FOR A ROADMAP TO ENVIRONMENTAL SUSTAINABILITY IN THE NIGER DELTA
A KEYNOTE ADDRESS BY THE HONOURABLE MINISTER
MINISTRY OF NIGER DELTA AFFAIRS, FEDERAL REPUBLIC OF NIGERIA
ELDER GODSDAY ORUBEBE
AT THE ACADEMIA FOR GREEN AFRICA INTERNATIONAL CONFERENCE
HELD ON 19TH OCTOBER 2010 AT BENSON IDAHOSA UNIVERSITY,
BENIN CITY, EDO STATE

It is my pleasure and privilege to deliver this keynote address at the opening of the Academia For Green Africa (AFGA) Conference 2010 under the auspices of the initiators of the AFGA—Benson Idahosa University and Envirofly Consulting UK Limited. First of all, I note that in line with current global trends, the theme of the conference, "**TOWARDS A GREENER AFRICA: THE ROLE OF ACADEMIC INSTITUTION IN CAPACITY BUILDING**", is most appropriate and I wish to specially commend Benson Idahosa University and Envirofly Consulting UK Limited for organizing this epoch making event of bringing together African and international scholars, corporate leaders, policy makers and other stakeholders to address the issues of environmental sustainability with particular focus on the impacts or threats of global warming and climate change as it concerns economic and social development in Sub-Saharan Africa.

2. No doubt, one of the most topical global environmental issues, today, is environmental sustainability and climate change. There is overwhelming consensus in the scientific community and the global political leadership that climate change is real and mostly human induced. A general overview of the situation of Sub-Saharan Africa paints a very disturbing picture. It is one of the most threatened parts of the world by the twin devastating effects of rapidly increasing environmental degradation and climate change. The desertification process is increasing; more areas are now prone to longer periods of drought, higher and more devastating incidences of flooding and erosion, environmental pollution from industrial and domestic activities, water crisis, etc. There is also a rising level of poverty, hunger and food insecurity in the region. There are more environmental refugees in Sub-Saharan Africa than those displaced through conflicts. Scholars are predicting that about 50 million people worldwide will be displaced by this year 2010 as a result of rising sea levels, desertification, dried up aquifers, weather-induced flooding and other serious environmental changes with a greater number from Sub-Saharan Africa.

3. The environmental issues in Nigeria represent in microcosm what is happening in Sub-Saharan Africa. Nigeria has five distinct ecological zones, namely the Sudan-Sahel, Guinea Savanna, Derived Savanna, Moist Rain Forest and the Swamp/Mangrove Ecosystems, beginning from the drier northern part of the country to the delta environment in the southern most part of the country. The nation is also blessed with abundant mineral resources, including oil and gas, with a largely semi-urban/rural agrarian based population.

4. My keynote address reflects on environmental sustainability of the Niger Delta region. The reason for this emphasis is because, as you are all aware, the Niger Delta is the main source of revenue to government in Nigeria because of the nation's high dependence on oil and gas economy. Exploitation and production activities in the past have resulted in serious degradation of the environment of the Niger Delta. This has also brought some serious consequences and negative impacts on the environment, health and general well being of the peoples of the Niger Delta of Nigeria. Policy failures and lack of commitment by the oil companies have been largely responsible for this situation which until recently led to an atmosphere of grave insecurity as a result of militant activities in response to the neglect of the region.

5. The Ministry of Niger Delta Affairs was established in December 2008 at the height of the insecurity and environmental crisis in that region with the mandate for a coordinated and holistic development of the Niger Delta. Several programmes have been initiated by the Federal Government of Nigeria since then. The programmes include the Amnesty and infrastructural and environmental development programmes.

6. Within the context of climate change, the Niger Delta by virtue of its geographical location is particularly vulnerable to the impact of climate change on the environment. Apart from the increased amount of rainfall which has given rise to flooding and erosion, several settlements have been displaced as a result of sea level rise and incursions along the 540km coastline of the Niger Delta Region out of Nigeria's 850km coastline, resulting in environmental crises. The crisis has been aggravated by the poverty levels and lack of development. Furthermore, as the oil producing region in the country, emission of green house gases from gas flaring and other industrial activities has direct effect on global warming.

7. The Ministry of Niger Delta Affairs is committed to reversing the trend of environmental degradation in the region as well as coordinating a holistic development of that region. The environmental programmes of the Ministry include the following:

    (i)   Land reclamation and shoreline protection;
    (ii)   Dredging/canalization of inland water ways/rivers;
    (iii) Rehabilitation of degraded ecosystems through tree planting as carbon sink and for other social and economic benefits;
    (iv)   Remediation and rehabilitation of oil impacted sites;
    (v)   Environmental Impact Assessment of proposed projects and Impact Mitigation Monitoring of major on-going projects;
    (vi)   Environment sensitive index mapping of the Niger Delta Region;
    (vii) Niger Delta State of the Environment Reporting; and
    (viii) Pipeline and oil facilities surveillance and monitoring.

8. I am happy to note that AFGA is focused at bridging the environmental information gap in Africa's academia and build general awareness on environmental issues. I am also enthused by the knowledge that the initial focus which is on human resources development, environmental research, information retrieval and adaptable projects, will help to obviate many of the environmental problems in Sub-Saharan Africa. The Ministry of Niger Delta Affairs is therefore committed to supporting these initiatives.

9. Distinguished participants, as you are aware, the Niger Delta is one of the most vulnerable coastal environments in the world, along with the Mekong, Irrawaddy, Ganges Brahmaputra and the Nile Deltas. It is with this in mind that I urge you to also address the peculiar environmental issues of the Niger Delta and recommend strategies and necessary synergies for the attainment of environmental sustainability. I would like to therefore bring up some specific issues, amongst others, for your scientific consideration:

    (a) The development of an effective national and regional coordinating mechanism for the collation, analysis, interpretation and dissemination of information based on time series environmental indicators for proper understanding of environmental problems, causes and necessary mitigation and adaptation measures;

    (b) Appropriate planning regulations and enforcement mechanisms and regional cooperation;

    (c) Ecosystems and biodiversity management;

    (d) Oil pollution and gas utilisation;

    (e) Environmental Sanitation and integrated waste management; and

    (f) Energy efficiency technology and use.

10. In conclusion you will agree with me that addressing decisively environmental issues in Sub-Saharan Africa in a coordinated manner with all stakeholders on board is the key to ensuring sustainable development, poverty eradication and economic growth in the continent. Judging from the calibre of persons presenting papers, the various backgrounds of participants at this conference and with the topics slated for discussion, I am full of hope and expectations that the outcome of the conference will meet the expectations of the conveners towards attainment of environmental sustainability and development in Sub-Saharan Africa as well as assist my ministry in promoting environmental sustainability in the Niger Delta of Nigeria.

11. Once again, I thank you for inviting me to address this conference and wish you fruitful deliberations.

ELDER GODSDAY ORUBEBE
HONOURABLE MINISTER OF NIGER DELTA AFFAIRS, FEDERAL REPUBLIC OF NIGERIA

5.0 KEYNOTE ADDRESS BY
THE HONOURABLE MINISTER,
FEDERAL MINISTRY OF ENVIRONMENT, FEDERAL REPUBLIC OF NIGERIA
MR. JOHN ODEY
AT
THE FIRST ACADEMIA FOR GREEN AFRICA (AFGA)
INTERNATIONAL CONFERENCE
ON
"TOWARDS A GREENER AFRICA: THE ROLE OF ACADEMIC INSTITUTION IN CAPACITY BUILDING"
AT
BENSON IDAHOSA UNIVERSITY,
BENIN CITY, 19TH-21ST OCTOBER 2010

It is indeed my pleasure to address you on the occasion of the 1st Academia for Green Africa (AFGA) International Conference holding today at the Benson Idahosa University, Benin City.

2. The Federal Ministry of Environment of Nigeria particularly welcomes this initiative and considers the theme of this Conference, which is, **"Towards a Greener Africa: The Role of Academic Institutions in Capacity Building", very apt.** I commend the organizers and hosts of the programme. I have no doubt that the objective of this first Conference which is, **"to identify and enhance the role of African Academic institutions in achieving a greener Africa,"** will be achieved.

3. Nigeria, like other countries in Africa is faced with the task of promoting economic development that meets the needs of its population, while ensuring that the environmental ecosystem on which the people rely for their livelihood is not endangered or destroyed in the process.

4. The Federal Government of Nigeria appreciates the place of the Environment as the fabric, pivot and resource base on which all development efforts depend. Government recognises the tremendous importance of natural resources to the existence and survival of humanity and is determined that maximum benefits are derived from our natural environment while protecting and conserving it for the benefit of future generations.

Challenges Confronting the Nigeria Environment

5. The challenges facing the Nigeria environment include: Deforestation, Erosion and Flooding, Desertification, Drought, Oil Spillage and Gas flaring, Pollution from various sources including Greenhouse Gases (GHGs) and Carbon emissions, Effects of climate change.

Implementation of Intervention Strategies

6. Nigeria is determined to tackle the environmental challenges that pose a threat to our socio-economic development. In this regard, relevant strategies have been adopted and are being executed aggressively and effectively by Government. These include as follows:

I. The execution of a Presidential Initiative on Afforestation to combat desertification through intensive afforestation projects in each of the eleven frontline states of the nation.

II. The establishment of relevant institutional mechanisms for environmental management, compliance monitoring and enforcement of environmental laws, guidelines and regulations. These include:

- The establishment of a Special Climate Change Unit to enhance the effective implementation of the United Nations Framework Convention on Climate Change (UNFCCC) and the Kyoto Protocol to which Nigeria is a party.
- The establishment of the National Environmental Standards and Regulations Enforcement Agency (NESREA) and the National Oil Spill Detection and Regulatory Agency (NOSDRA).
- Strengthening of existing agencies, including the Forestry Research Institute of Nigeria (FRIN) and the National Parks Service and execution of relevant policies and action plans for effective environmental management in all priority environmental areas, including erosion, flood control and coastal zone management, forestry, biodiversity management, chemical and solid waste management, renewable energy, climate change adaptation and mitigation, biosafety and environmental impact assessment.
- Execution of relevant strategies for enhanced environmental education and awareness, stakeholders and community participation in environmental management.
- Cooperation with relevant agencies, friendly nations and the private sector in environmental initiatives and best practices.

7. Cross-cutting approaches being adopted in tackling the nation's environmental challenges include: Collaboration with relevant stakeholders in the development and execution of relevant policies, programmes and projects; Preparation of Pilot Environmental Projects at the three tiers of governance.

8. Climate Change is currently the most pronounced global challenge besides HIV/AIDS, hunger and diseases. Nigeria is gradually taking a leading role in Africa on mitigation and adaptation. Some of the activities of Nigeria at the local, regional and global levels include:

- Participation in the Reduced Emissions from Deforestation and Degradation (REDD+) Programme; Creation of Sinks for Greenhouse Gases through the Multilateral Agreement in the Green Wall Sahara Project, mass tree planting and campaigns; Cutting down on gas flaring to reduce the emission of methane and petroleum associated gases; Participation in the implementation of Clean Development Mechanism (CDM), through execution of Kwale Gas Gathering Project, Pan African Gas Gathering Project and Sare-80 Wood Store Project.
- Adoption of appropriate strategies for effective public-private sector partnership in the conceptualization and development of specific intervention projects for environmental management.

9. Complementarily, in pursuance of Government commitment to Climate Change Adaptation, Nigeria is committed to the following three major efforts:

- Development of National Adaptation Strategy and Plan of Action (NASPA)
- Establishment of a Ministerial Steering Committee on the Kyoto Protocol Adaptation Fund.
- Collaboration with key Stakeholders on climate change adaptation issues.

10. In the area of Pollution Control, the Ministry's interventions include:

- The implementation of an Integrated Solid Wastes Management Programme through public, private partnership; community-based pilot municipal solid waste management facilities in some cities.
- Hazardous Hospital Waste—provision of gas fired incinerators to four (4) teaching hospitals.
- Construction of Prototype Gas-Phase Chemical Reduction (GPCR) Plant for environmentally sound destruction and reverse transformation of persistent organic pollutants (POPs) waste in Minna, Niger State.

In recognition of the hazards of chemical products waste, the Ministry has adopted a more professional approach in the management of chemicals by establishing Chemical Emergency Response Centers across the country under the African Stockpile Programme (ASP), among which are:

- Establishment of Ozone Technology Village in Ikenne Local Government of Ogun State, with the objective of creating an ozone friendly technology village for the environmentally friendly equipment and technology development.
- Construction of a Prototype Gas-Phase Chemical Reduction (GPCR) Plant for an Environmentally Sound destruction and Reverse Transformation of Persistent Organic Pollutants (POPs) Waste in Minna, Niger State; aimed at the development of a POPs Management Demonstration Centre equipped with a prototype Gas-Phase Chemical Reduction Plant for an environmentally sound destruction and reversal of persistent organic pollutants (POPs) waste.
- Inventory of National Ambient Air Quality in some selected cities in Nigeria, namely: Lagos, Port-Harcourt, Kano, Aba, Abuja and Maiduguri for the establishment of National Air Quality Monitoring and Management Stations; with the objective to carry out inventory, access data and produce a national air quality management policy document for the nation.

Gender in Environment

11. In recognition of the importance of gender in environment for sustainable socio-economic development, Nigeria has intensified effort in mainstreaming gender in environment policies and programmes. These include the establishment of a Ministerial Gender Board comprising major development sectors, approval for the execution of a Gender in Environment Policy and Implementation Manual suitable for trainers, capacity building as well as advocacy and sensitization programmes in various communities and by state governments.

Expectations

12. I have no doubt that this Conference will be holistic in its approach as it meaningfully addresses the economic, political, social, legal, historical and the scientific questions, as well as appropriate response to the threats of global warming and climate change, particularly with respect to economic and social development in Sub-Saharan Africa.

The Academia And Environment

13. Distinguished ladies and gentlemen of the letters, in all the focal areas mentioned above, some of the roles expected of the academia include but not limited to the following; need to:

    - raise awareness on the impacts of climate change from the rural community level to the policymakers' level;
    - undertake systematic observation and research activities with respect to inventory of greenhouse gases in the Nigeria atmosphere;
    - assess more comprehensively, the country's vulnerability to, and design appropriate adaptation strategies to the impacts of climate change and promote regional cooperation and coordinated solutions to jointly address a common threat;
    - participate at international fora not only for the continued negotiations to further elaborate the implementation of, but also to access multi-lateral funds available for the implementation of the Convention and the Protocol;
    - collaborate with governments, civil society organizations, international bodies and donors to use research outputs and plan to support adaptation by African/Nigerian populace, and to prepare for extreme events;
    - Ensure that the local end-users are able to access and use climate forecasts and other assessments results for planning their activities.

14. We are convinced that this International Conference will result in appropriate partnership and network arrangements among other strategies, to strengthen capacities across our great continent through enhanced commitment and cooperation among the academia and relevant sectors.

15. On our part, the Federal Ministry of Environment of Nigeria, in line with Government policy, mission and vision, will continue to collaborate with the academia as appropriate in the concerted march to achieve environmentally sound and sustainable socio-economic development as well as relevant regional and global obligations for the common good of our great continent and this **"One Planet."**

16. I wish you successful deliberations and look forward to your recommendations for appropriate action.

Thank you for your attention

6.0

**TECHNICAL PAPER
ON CLIMATE CHANGE SUBMITTED
BY
THE HONOURABLE MINISTER,
FEDERAL MINISTRY OF ENVIRONMENT,
MR. JOHN ODEY,
TO
THE FIRST ACADEMIA FOR GREEN AFRICA (AGFA) INTERNATIONAL
CONFERENCE,
ON
"TOWARDS A GREENER AFRICA: THE ROLE OF ACADEMIC INSTITUTION IN
CAPACITY BUILDING"
AT
BENSON IDAHOSA UNIVERSITY, BENIN CITY,
19th - 21st OCTOBER, 2010**

Protocol

In the recent years, international concern on the issues of climate change and global warming has heightened. Scientific evidences have proven beyond reasonable doubt that climate change is affecting patterns of life and general living conditions of people around the world. It is affecting the availability of water, food production, health, cultures, economic, well-being, recreation and tourism among others. These are some of the reasons why issues of climate change have continuously been placed firmly on the front burner in very many global and national discuss.

Proven scientific evidences have shown that millions of people, especially in developing countries, including Nigeria, will continue to be confined to perpetual poverty and precarious living conditions due to food and water shortages, coastal flooding and extreme changes in weather pattern as a result of global warming which are warning signals. Reminding us all of the recent flooding in Sokoto, Edo, Adamawa to mention but a few may rather be needless

The negative impacts of climate change such as temperature rise and attendant heat stress; erratic yet virtually unpredictable rainfall, sand storms, desertification, flooding, low agricultural yields, drying up of water bodies, e.g (Lake Chad), flooding, incidences of new diseases and disease vectors are real all around Nigeria.

Our climate is changing. Over the past century, average global temperatures have risen by 0.6° Celsius—**the 10 warmest years on record have all been since 1990**. There is a scientific consensus that this warming has been brought about by the increase in greenhouse gases in the atmosphere, which in turn has been caused by human activities—primarily the burning of fossil fuels and changes in land use.

Distinguished participants, may I remind you that all ambitious frameworks agreed by world leaders to reduce poverty and attain sustainable development, such as the Millennium Development Goals (MDGs) and the Seven Point Agenda, are at the danger of remaining a mirage due to retrogressive looming dangers of climate change. Although Nigeria contributes an insignificant amount of greenhouse gases emission to the global total, the country is among the most vulnerable to the impacts of climate change. This further call for urgency not only in the mitigation but more importantly also adaptation strategies in order to contain the threat posed to the country.

It is against this background that this meeting is considered significant and timely to brainstorm and come up with concise, practical and implementable recommendations that will assist in tackling the effects of climate change in the entire country as well as bringing out recommendations for the best practices for adapting to the menace. Participants should not only adopt a collective approach but also commit themselves to strategic actions aimed at reducing the vulnerability of their region against the looming dangers of climate change.

Addressing the climate change challenges must include both adaptation to address the inevitable, and mitigation to prevent the avoidable. Responses have to be developed in the face of uncertainties on the timing, location and severity of climate impacts. To respond to the adverse impacts of climate change, all vulnerable sectors must be identified and specific adaptation strategies should be developed for them. And this makes the case for vigorous investment in information and better understanding. Adaptation requires assessment of possible threats and opportunities arising from climate variability; and incorporation of the outcomes of such assessments into policy through the appropriate mechanisms.

Distinguished academia, time has come to take the bull by the horn. We need to raise awareness on the impacts of climate change from the rural community level to the policymakers' level. We need to undertake systematic observation and research activities with respect to inventory of greenhouse gases in the Nigeria atmosphere; we need to assess more comprehensively, the country's vulnerability to, and design appropriate adaptation strategies to the impacts of climate change and promote regional cooperation and coordinated solutions to jointly address a common threat. More importantly, we need to participate at international fora not only for the continued negotiations to further elaborate the implementation of, but also to access multi-lateral funds available for the implementation of the Convention and the Protocol.

Worthy of reiterating here is that the responsibility for action on the climate problem does not fall exclusively on the government. Admittedly, government has commenced actions in taking legislative or regulatory measures, prepare economic policies or incentives or dissuasive measures for economic operators. However, every sector of the society has important roles to play. Other private enterprises, producers, and consumers are encouraged to also stand out in the battle against greenhouse gases effects through insisting on sustainable and environmentally friendly product choices and production techniques.

In recognition of the impacts of climate change on our socio-economic development as a developing country, Nigerian government, has been committed to the issue of climate change and thus put in place strategic action plans, policies, regulations and other meetings as initiatives to address the challenges of climate change and to also draw the opportunities therein. One of them is the establishment of the Special Climate Change Unit.

The Ministry through the Special Climate Change Unit implements these policies by working in collaboration with other relevant government organizations, non-governmental organizations, academia and private sector through an Inter-ministerial Committee on Climate Change (ICCC), which is the policy advisory organ for government under the Chairmanship of the Federal Ministry of Environment while the National Committee on Climate Change (NCCC), serves as the technical advisory organ. The committee meets regularly on quarterly basis and on ad-hoc basis to review polices on climate change, advice government on appropriate actions, and put up Nigeria's position at meetings where climate change issues are being discussed or negotiated.

The Ministry collaborates with the National Linkage Centre on Climate Change and Freshwater Resources situated in the Federal University of Technology, Minna in areas of training, research and data gathering on climate change. The Ministry also collaborates bilaterally and multilaterally with UNIDO, UNDP, UNEP, as UN organizations. Locally, she collaborates with NGO's, CBO, Faith Institutions, among which are Heinrich Boll Foundation (HBF), Nigeria Climate Action Network (NigeriaCAN) and others. For instance the Federal Ministry of Environment in collaboration with Nigeria Environmental Study/Action Team (NEST), through funding from the Canadian International Development Agency (CIDA) and also in collaboration with Canadian Organization, Global Change Strategies Incorporated, (GCSI), has from 2001—2007 embarked on a project named the Canada-Nigeria Climate Change Capacity Development (CN-CCCD) project aimed at public awareness raising and capacity building on climate change through a series of workshop, consultation and awards to intermediary organizations and research institutions. Presently, there is also the on-going "Building Nigeria's Response to Climate Change" which is a second phase of the above project.

There is Gender and Climate Change initiatives titled, "Women in Climate Change". This is a First Lady Initiative that pursues the plight of women and children, being the most vulnerable group of the population.

The Ministry of Environment has also flagged-up the National Strategy for Nationwide Awareness Campaign under a Public/Private Partnership Initiative that is aimed at public awareness creation.

In addition to the above, Mr. President has approved the establishment of an International Climate Change Centre to be known as "International Green-Hall of Fame" that will be used to honor individuals and groups that excel in activities that are aimed at combating climate change. It will also pursue carbon sequestration and tree planting.

The initiative by the Unit to see that all 36 states Ministry of Environment and FCT Environmental Protection Board establish a climate change desk has long been implemented.

Nigeria had inaugurated the NNPC-CDM Working Group. This Working Group is tasked with developing carbon market and finance in the oil and gas industry of the country.

Another major policy goal of the Government is to ensure sustainable use of our forests. Towards these ends, Government has embarked on elaboration of national tropical forest action plan, launching of extensive reforestation programmes in the southern part of the country and afforestation programmes in the northern part of the country. Such programmes include community-based tree planting programmes, tightening control of fuel wood extraction from reserves and development of more efficient wood stove. The goal is not only to protect the sinks for carbon dioxide as one of the means of mitigating climate change but also to safeguard our biological diversities and reverse the ecological status of the area to a more humane one.

Nigeria plans to embark on an initiative aimed at energy saving strategy and energy efficient system, contained in a CDM financed program, where all aspects of renewable energy initiatives, such as solar, wind, biomass, geothermal, and small hydro will be encouraged nationwide. This will increase the bottom-line of bankable Clean Development Mechanism (CDM) projects and enhance our national sustainable livelihood.

More recently, government efforts to combat climate change include:

- The passage of a Bill in the National Assembly seeking to establish a National Climate Change Commission. Currently, the Bill is at the harmonization stage between the two chambers of the legislature. When concluded, the new governance structure is expected to warehouse the multi-sector concerns and streamline the various approaches (mitigation and/or adaptation) to address the problem. It will also provide a focus for engaging bilaterally with international groups as well as meeting international obligations;
- Nigeria successfully led a large team of National experts, policy makers and related stakeholders to the United Nations Climate Change Conference held in Copenhagen in December 2009;
- Establishment Of Databank Management System for the Generation and documentation of Geo-Spatial Information and National Reference Emission Level;
- Nigeria is in the process of developing her Strategic Framework for voluntary National Appropriate mitigation Action (NAMA) for attracting funds under the COP 15 political agreement to enhance our domestic actions to reduce the country's emission level;
- Nigeria is in the final stage in the development of her National Adaptation Strategy and Plan of Action. This is with the major objective of providing an integrated strategy and action plan designed to reduce the impacts of climate change through adaptation measures, to be put in place by governments, civil societies, and the private sector, including measures that will:
- Increase the flexibility and resilience of infrastructure, agriculture and natural resources.
- Improve awareness and preparedness for climate change impacts on health, human security and livelihoods.
- Integrate adaptation into national and sectoral planning, and into State and Local Government Authority (LGA) planning.
- Initiation of Nigeria's National Climate Change Policy and Response Strategy which became necessary in order to effectively deal with already manifested and potential problems associated with climate change; involving understanding our vulnerabilities and analysis critical to developing appropriate adaptation programmes, whence, encompassing assessment of exposure, adaptive capacity and sensitivity;

- Development of National Green House Gas Inventory; with the main aim to ascertain the country's precise or near precise individual and sectoral greenhouse emission quantity and level;
- Inauguration of the National Technical Committee for REDD and REDD +; Reducing Emissions from Deforestation and Forest Degradation and contribute to conservation, sustainable management of forests and enhancement of forest carbon stocks. This project has the potential to deliver significant social and environmental co-benefits. Although many have also highlighted the serious risks, particularly for Indigenous Peoples and other forest-dependent communities, the project aims to define and build support for a higher level of social and environmental performance;
- Nigeria just commenced the project concept initiation for the development of her National Climate Change Trust Fund.

The Academia and Climate Change

The academia being the engine room for the conceptualization and development of strategic framework for all issues affecting human existence has a major role to play in the development of adaptation and mitigation strategies of climate change. Some of their expected strategic roles are itemized below:

- Need for African scientists and research organizations to contribute to adaptive capacity by carrying out impact assessment and examination of options for adaptation (including linkages and co-ordination between organizations);
- To collaborate with governments, civil society organizations, international bodies and donors to use research outputs and plan to support adaptation by African/Nigerian populace, and to prepare for extreme events;
- To collaborate with international scientific community to support, strengthen and complement the work of African scientists and governments.
- Ensure that the local end-users are able to access and use climate forecasts and other assessments results for planning their activities;
- Participating in orientating and empowering communities to better understand the issue on climate change and manage their risks.
- Participate in assessing and feed collating feedback from communities in order to provide useful linkage information on climate change and poverty.
- The academia would also need full access to climate relevant information systems;
- Supporting adaptation by rural and urban people, particularly the most vulnerable;
- Mainstreaming climate change studies into the academic calendar.
- Integration of impacts into macroeconomic projections. The rate and pattern of economic growth is a critical element of poverty eradication, and climatic factors can have a powerful bearing on both. Integration will prevent Climate Change diverting limited resources into disaster relief and recovery activities and away from long-term development priorities. The national budget process should be the key process to identify climate change risks and to incorporate risk management so as to provide sufficient flexibility in the face of uncertainty.

I am optimistic that at the end of this conference, you will be able to come up with a strong position on the way forward. Moreover, all participants here present as well as the populace reading the communiqué thereafter would be better equipped on all issues relating to climate change especially as it affects us as citizens of Africa and the world at large.

I wish you all fruitful deliberations.

Thank you.

7.0 SCIENTIFIC PAPER PRESENTATIONS

7.1: Economic Development and Environmental Degradation: Testing for the Existence of the Environmental Kuznets Curve in Sub-Saharan Africa.

JEL Classification: Q56, Q52, P52

Sam Guobadia[1]
Leonard Aisien[2]

Abstract

Empirical evidence from many industrialized economies seems to suggest that there is a positive relationship between economic growth and eco-efficient production. According to Brock and Taylor (2004) and others, environmental sustainability is more difficult to achieve at low levels of income than at high levels of income. This describes the income pollution relation of the popular Environmental Kuznets Curve (EKC) which has been shown to be valid in the developed economies. This paper examines the validity of the EKC in selected sub-Sahara African Countries using time series from 1960-2005. The model was specified in quadratic form in line with existing literature and the error correction modelling technique was adopted in the estimation in order to examine both the short run and long run characteristics of the model. The variables utilized in the paper include GDP per capita (current US$) and Carbon dioxide (CO2) emission for five selected Sub-Sahara African Countries. On the basis of the empirical results, the paper opines that the EKC hypothesis is valid for Sub-Saharan Africa. The computed Income Turning Points (ITP's) for the countries were found to be relatively high compared to their current incomes. This shows that most Sub-Sahara African countries are yet to reach their income turning points hence they are witnessing economic growth accompanied by environmental degradation. Implied in the results is the need for collaborative efforts between the developed and developing countries in order to attain eco-efficient production and greener economies worldwide.

Key words: Environmental Kuznets Curve, income turning point, Carbon dioxide (CO2), pollution, eco-efficiency, green economy

[1] Sam Guobadia, Ph. D, is Senior Lecturer and Acting Head of Department, Economics, Banking and Finance, Benson Idahosa University, Benin City, Nigeria. Dr. Guobadia is also Director of the Consultancy Services and Head of the Academia for Green Africa (AFGA) in the same institution.

[2] Mr. Leonard Aisien is a Lecturer in the Department of Economics, Banking and Finance and a Member of AFGA Board, Benson Idahosa University, Benin City, Nigeria

Terminology and Abbreviations

Greening: Economic activities that are environmentally sustainable and are socially responsible.

Eco-efficient production: Eco-friendly production; acquisition of new technology that reduces GHGE and encourages the development & use of renewable energy.

Economic development: Traditionally defined as industrialization with the associated rise in environmental degradation, green house gas emissions, and global warming.

GHGE: Green house gas emission

ITP (Income Turning Point): The point countries are rich enough to be able to afford eco-efficient production technology/techniques.

IPR (Income Pollution Relation): Depiction of the EKC

CDM: Clean Development Measurement under the Kyoto Protocol.

IPC: Income per capita

EKC: Environmental Kuznets Curve

SSA: Sub-Saharan Africa

1   Introduction

The drive to achieve greener economies has been receiving worldwide boost in recent times because of the realization and importance of the interdependence between the natural ecosystem and human economies. A *green* economy is defined as a combination of economic activities that are environmentally sustainable and socially responsible. The focus of the economy include reducing energy consumption, reducing pollution to minimum level, substituting fossil fuel with renewable energy, promoting low carbon energy production and emission as well as the reduction in deforestation.

The drive to build greener economies began in the advanced economies in the 80's. This drive has been extended to the developing economies. However, the most pressing economic issues facing the developed and developing countries may not necessarily be the same. The industrialized countries which depended on heavy polluting industries for their wealth in the course of economic development now fear that uncontrolled economic activities in the developing countries (massive destruction of tropical rain forest, soil destruction, gas flaring, oil spills, etc) would lead to environmental disaster. Thus their current focus is how to provide leadership in the transformation to a greener global economy. Through innovative environmental and infrastructural policies, they are currently driving demand for green products. On the other hand, the developing countries are preoccupied with the issues of industrialization, economic development and building their own wealth. The primary goal of these countries is support and sustainability of their growing population. Achieving development through industrialization is their current priority with less emphasis on environmental sustainability. This is even more so as they believe that the current global environmental damage is attributable to the industrial activities of the industrialized countries. It is quite obvious why little efforts have been made in mainstreaming climate change or global warming issues into economic development planning in Sub-Saharan Africa. Going by the traditional development theories, it will be difficult, if not impossible, for Sub-Sahara African countries to pursue environmental sustainability agenda at the expense of of higher economic growth. Clearly, there exists a conflict between development and environmental conservation for these countries.

Empirical evidence for several industrialized countries has shown that at the early stage of economic growth, environmental degradation and pollution increase. But beyond some level of income per capita the trend reverses. This gives rise to the inverted U-shaped function popularly referred to as the Environmental Kuznets Curve (EKC). The concept of the Environmental Kuznets Curve emerged in the early 1990's stemming from the works of Grossman and Knueger (1991) and Bandyopadhyay (1992). The theme was further popularized by the World Bank in its World development report 1992 (IBRD, 1992). In the report, it was argued that "as income rises, the demand for improvement in environmental quality will increase as well as resources for investment" (IBRD, 1992). This position was also emphasized in Beckerman (1992); ". . . although economic growth usually leads to environmental degradation in the early stage of the process, in the end, the best and probably the only way to attain a decent environment in most countries is to become rich".

The Environmental Kuznets curve is currently one of the most commonly used concept to analyze the pollution income relation. But all the empirical evidence to support the existence of the Environmental Kuznets Curve is based on data from the developed economies except Desgupta, Laplanta, Wang and Wheeler (2002). A search of the extant literature has not shown any study devoted specifically to Sub-Saharan Africa. This thus raises the question whether or not the Environmental Kuznets curve is applicable to the Sub-Saharan Africa, and if it does, what constitutes the Income Turning Points for the countries in the region.

Therefore, the main objective of this study is to examine the applicability of the Environmental Kuznets Curve in Carbon dioxide (CO2) emission in Sub-Saharan Africa based on time series data for selected countries in the region. The study relies on the use co-integration and error correction modeling techniques to carry out the investigation. Another objective of the study is to contribute to filling the research gap on the subject matter in Sub-Saharan Africa; collecting and developing uniquely African environmental data and undertaking empirical studies that will assist policy.

The magnitude of atmospheric concentration of carbon dioxide (CO_2) is one reason it is selected as the major variable for the study. CO_2 among other green house gasses contribute immensely to the phenomenon of global warming. The verification of the existence of EKC in Sub Saharan Africa is very crucial at this point in time as it will help in guiding policy as counties strive toward achieving a greener global economy. Its existence would mean that growth enhancing policies would have to be combined with environmental sustainability programmes at the early stages of economic development of the region. On the other hand, if it doesn't apply, it means that economic growth will continuously result in environmental degradation.

The paper is structured into five sections. Following section one, section two covers the review of relevant literature. While section three deals with the theoretical issues, section four provides important empirical analysis. The study concludes with section five, providing a list of recommendations that will guide environmental policy making in the region.

2   Literature Review

Although the concept of the Environmental Kuznets Curve is a recent phenomenon, however, there is a vast literature for the use of the concept (Kuznets, 1955). Grossman and Krueger (1991) pioneered its application to environmental issues. The study investigated the environmental impact of a North America Free Trade agreement. An environmental Kuznets curve (EKC) was estimated for Sulphur dioxide (SO_2) and Suspended Particles Matters (SPM). An inverted U-shaped Environmental Kuznets curve for both pollutants was confirmed. The income turning point was estimated to be around $4000-$5000 for SO_2, while the concentration of Suspended Particles appeared to decline even at a low income levels.

Moomaw and Unrah (1997) and Unruh and Moomaw (1998), in their studies, find evidence that Carbon dioxide (CO_2) emission trajectories of Sixteen OECD Countries follows an inverted U shaped curve with respect to time and not income. The turning point for all countries under study occurred around 1973 following the first worldwide oil price hikes.

Other studies which have investigated the existence of the environmental Kuznets Curve in the OECD countries include Selden and Song (1994), Dijkgraaf and Vollebergh (1998), Stern (2002) and Vorgetegt and Egli (2005). Selden and Song (1994) estimated an income turning point of $10,391, while Dijkgraaf and Vollebergh (1998) estimated a turning point of 54% of maximum GDP in the sample. Stern (2002), used econometric model to decompose sulfur emission in 64 countries for the period between 1973 and 1990. The result also shows evidence of inverted U-shaped curve. Vorgetegt and Egli (2005) find evidence of the existence of the environmental Kuznets curve based on time series data for Germany. The study investigated the relationship between several pollutants and income. The pollutants were Sulphur dioxide (SO_2), Nitrogen oxide (NO_2), Carbon dioxide (CO_2), Carbon monoxide (CO), Ammonia (NH_3), Methane (CH_4), Particulate matter (PM) and non methane volatile organic compound ((NMVOC) for the period 1966-2002.

The literature contains many studies testing the validity of EKC based on data from the advanced countries. However, Desgupta, Laplanta, Wang and Wheeler (2002), presented evidence that environmental improvement is possible in developing countries and that peak levels of environmental degradation will be lower in countries that are able to achieve sensible steady growth. They presented data that shows decline in various pollutants in developing countries overtime. This study has shown that the inverted U shaped EKC is not only applicable to the industrialized countries alone but could also be applicable SSA countries.

Apart from the above studies which confirmed the inverted U shaped of the environmental Kuznets curve, other studies such as Toras and Boyce (1998) and Partly (1994) find no relationship between development and environmental quality. They thus doubt the existence of the environmental Kuznets curve or its inverted U shaped.

In sum, the majority of the studies have confirmed the existence of the environmental Kuznets curve and its shape, with different income turning points in relation to various pollutants. The motivation for this study stems from the fact that no such data exists for SSA. The countries selected for this study include Nigeria, Kenya, Mauritius, South Africa and Seychelles for the research investigation and the applicability of the EKC.

3   Theoretical Framework

The theoretical underpinning of the inverted U-shaped environmental Kuznets curve is the structural and technological changes that come with economic growth. "if there were no changes in the structure of technology of the economy, pure growth in the scale of the economy would result in a proportional growth in pollution and other environmental impacts" (Stern, 2003). This is referred to as the scale effect. The traditional theories of development in theorizing the impact of development on the environment often hold technology constant. Thus the views of the theorists are based on the scale effect alone. However, according to Panayotou (1993), one of the proponents of the environmental Kuznets curve, "at higher levels of economic development, structural change towards information intensive industries and services, coupled with increased environmental awareness, enforcement of environmental regulations, better technology and higher environmental expenditure, results in leveling off and gradual decline of environmental degradation".

The theoretical literature on the environmental Kuznets curve can be classified in two groups. The first group stresses shift in the production technology which differs in their pollution intensity as the main cause for the inverted U-shaped environmental Kuznets curve. The set of theories sees pollution as a product of production activities alone. As the economy becomes wealthy, more funds are available to acquire improved technology that is environmentally friendly. At this point also, there is a transition from manufacturing intensive production to a service intensive production. This shift in production technology will lead to a turning point in the environmental degradation.

The second set of theories stress that the shape of the curve results from the explicitly modeled abatement of gross pollution. These theories postulated that apart from consumption and investment in capital (human and physical), there is another economic activity known as environmental effort. Prominent among the theorists is Andreoni and Levinson (2001). In their model, they gave two conditions that must hold for the environmental Kuznets curve to occur:

i.   The marginal willingness to pay to clean up the last speck of pollution does not get to zero as income approach infinity
ii.   There must be increasing returns to scale in abatement.

They hypothesized that utility depends positively on consumption and negatively on pollution. This is specified as $U = U(C,P)$ where U, C and P respectively describes utility, consumption and pollution. Furthermore, pollution is a function of consumption and environmental effort, stated as $P = C - B$ (C,E); $B (C, E)$ is environmental effort, *i.e.*, abatement technology. It follows that pollution increases with consumption and decreases with increase in abatement technology. There is also the budget constraint, given as $M = C + E$; M is available resources or income.

This model parameterized $U(C, P) = C - Zp$, with $z = 1$ and $B(C, E) = C^\alpha E^\beta$ and thus shows that the environmental Kuznets curve will result provided $\alpha + \beta > 1$. This can be shown by inspecting the pollution function in terms of M:

$$P = C - C^\alpha E^\beta \qquad\qquad (1)$$

Multiplying through by M

$$P(M) = CM - (CM)^\alpha (EM)^\beta \qquad\qquad (2)$$

$$P(M) = CM - C^\alpha M^\alpha E^\beta M^\beta$$

$$P(M) = CM - (C^\alpha E^\beta)M^{\alpha + \beta} \qquad\qquad (3)$$

Equation 3 implies that $P(M)$ is concave in M provided $\alpha + \beta > 1$. Hence increasing returns to scale in abatement (defined by $\alpha + \beta > 1$) represent a necessary condition for the existence of an environmental Kuznets curve.

Thus, the proximate factors that may explain the existence of the environmental Kuznets curve include:

i.   Changes in the scale of production which implies expansion in production at a given factor—input ratio (the scale effect).
ii.   Improvement in the state of technology
iii. Changes in input mix. This involves the substitution of less environmentally damaging inputs for more damaging inputs.

These proximate factors can be driven by improved education, sound government environmental regulations and increased awareness on environmental issues.

4   Model Specification

Stemming from available literature on the environmental Kuznets curve, the following model is specified for income pollution relation:

$$P_t = \Psi_0 + \Psi_1 Y_t + \Psi_2 Y^2 + e_t \qquad (4)$$

Where:

P = Pollution indicator (CO_2 Emission)

Y = Income level (measured by per capita income in US$)

e = normally distributed error term

t = time index

An environmental Kuznets curve (inverted U Shaped) will result if:

$$\Psi_1 > 0 \ and \ \Psi_2 < 0$$

The Income Turning Point (ITP) which is the income level at which the environmental pollution begins to decline can be obtained by setting the first derivative of equation (4) with respect to income to zero. This will yield:

$$Y_t = -\Psi_1 / 2\Psi_2$$

In order to examine the time series properties of date used and also to distinguish between a long term income emission relationship and short term disturbance from the long term equilibrium path, an Error Correction Model (ECM) is specified. The ECM for this study is therefore specified as:

$$\Delta P = \psi_0 + \psi_1 \Delta Y_t + \psi_2 \Delta Y_t^2 + \psi_3(P_{t-1} - \psi_0 - \psi_1 Y_{t-1} - \psi_2 Y_{t-1}^2) + e_t \qquad (5)$$

Δ denote a variable's first difference while the expression $(P_{t-1} - \psi_0 - \psi_1 \Delta Y_{t-1} - \psi_2 \Delta Y_{t-1}^2)$ is the error correction term which denotes the deviation from the long term equilibrium. It coincides with the one period lagged residuals of equation (4), e_{t-1}.

For the inverted U shape to occur for the curve, the coefficients of ΔY_t must be positive, while ΔY_t^2 must be negative (ie $\psi_1 > 0$; $\psi_2 < 0$). Also, the coefficient of e_{t-1} must be negative.

Estimation methods

Most empirical studies on the validity of the environmental Kuznets curve adopted the cross country or panel data for their estimation. However, Roberts and Grimes (1997) argued that only single country studies could shed light on the validity of the environmental Kuznets curve hypothesis. Moreover, pooling countries in one panel can bias the estimation and therefore results may not be reliable. This distortion could be caused by the juxtaposition of different income emission relationships within the pooled countries. The criticism of the use of cross country or panel data in validating the existence of the EKC has been supported by Vencent (1997), Gallet (1999) and Dijkgraaf and vollebergh (2005).

To avoid the above problem, the study adopted a single country analysis based on time series data from five selected countries of Sub Sahara Africa (Nigeria, South Africa, Kenya, Senegal and Seychelles) for the period 1960-2005. The co integration and error correction modeling was utilized in estimating the EKC model. This is with the view of obtaining both the short run and long run relationship between pollution and per capita income. Also this will enable us examine the time series properties of data used. The data were obtained from the World Bank development indicators (WDI 2009).

5   Empirical Results and Analysis

(a) Time series properties of data (Stationarity test):

In testing for stationarty of the variables, the Dickey Fuller (DF) and Augmented Dickey Fuller (ADF) unit root test was adopted. The results for the five countries are given below:

Table I: Unit root test for variables from Nigeria

Variables	DF	ADF	Critical value
Y	- 6.9343	- 3.6401	- 2.9303
Y^2	- 6.5574	None	- 2.9303
CO2	- 2.0664	- 1.9552	- 2.9303
DCO2	- 7.0876	- 4.0258	- 2.9303

D = First order difference

Table II: unit root test for variables from Kenya

Variables	DF	ADF	Critical value
Y	- 0.9068	- 1.0499	- 2.9303
Y^2	- 0.4729	- 0.8247	- 2.9303
CO2	- 2.3250	- 2.1960	- 2.9303
DY	- 5.3056	- 3.4673	- 2.9303
DY^2	- 4.7668	- 3.2494	- 2.9303
DCO2	- 7.1055	- 5.2076	- 2.9303

D = First order difference

Table III: Unit root test for variables from Mauritius

Variables	DF	ADF	Critical value
Y	- 0.8599	- 1.0135	- 2.9287
Y^2	- 0.4282	- 0.7655	- 2.9287
CO2	- 2.3451	- 2.2056	- 2.9287
DY	- 5.3451	- 3.5130	- 2.9287
DY^2	- 4.8213	- 4.7165	- 2.9287
DCO2	- 7.2103	- 5.2807	- 2.9287

D = First order difference

Table IV: Unit root test for variables from South Africa

Variables	DF	ADF	Critical value
Y	- 0.2846	- 0.3071	- 2.9303
Y^2	- 0.2626	- 0.2626	- 2.9303
CO2	- 1.9073	- 1.8773	- 2.9303
DY	- 5.0523	- 3.0741	- 2.9303
DY^2	- 4.3568	- 1.7954	- 2.9303
DCO2	- 5.5168	- 3.9915	- 2.9303

D = First order difference

Table V: Unit root test for variables from Seychelles

Variables	DF	ADF	Critical value
Y	- 1.1148	- 1.3814	- 2.9303
Y2	- 1.1164	- 1.4827	- 2.9303
CO2	- 6.5574	- 4.6367	- 2.9303
DY	- 5.4073	- 5.3703	- 2.9303
DY2	- 4.2587	- 3.1350	- 2.9303

D = First order difference

(b) The Error Correction Results

Table VI: The parsimonious error correction representations for selected countries

	Nigeria	Kenya	Mauritius	South Africa	Seychelles
Constant	0.0292	0.0029	0.0793	0.0320	0.6819
Y	(0.7637)	(-0.5840)	(2.6728)	(0.6401)	(0.4954)
Y^2	0.4571	0.8668	0.1324	0.5313	0.1391
	(1.8312)	(3.1344)	(2.0597)	(1.9112)	(2.1512)
	-0.07181	-0.05441	-0.0152	-0.0722	-0.005147
	(-1.8312)	(-2.2024)	(-1.8845)	(-1.8562)	(-2.1521)

Ecm(-1)	-1.2000 (-6.4326)	-1.1757 (-8.3229)	-1.1000 (-5.4076)	-1.3100 (-5.4076)	-2.1521 (-1.9323)
R^2 F—STAT D. W	0.5542 19.5692 2.2483	0.6129 24.2250 2.3219	0.4143 6.3265 1.8439	0.4567 13.3314 1.8899	0.0371 1.5659 2.1383

From the above results, sign expectations of all the variables were met. The coefficients of determination were quit high for all the countries except Seychelles. Also the F statistics show a satisfactory performance.

The coefficients and t ratios of Y and Y^2 in the model for all the countries show that they are highly significant explanatory variables. The proper signs of the two variables in all the estimated results for all the countries confirm the validity of the Environmental Kuznets Curve for all the selected countries.

The results also clearly show a well defined error correction term (ECM) for all the selected countries. The effect of this disequilibrium error correction is not only large but also has the correct signs for all the countries. This revealed a high speed of adjustment from previous period disequilibrium. Thus, the model for all the countries under study is dynamically stable.

The computed income turning point for each country is shown in Table VII.

Table VII: Countries and their income turning points

country	Income turning point US $
Nigeria	3,182.70
Kenya	7,965.44
Mauritius	4,355.26
South Africa	3,679.36
Seychelles	13,512.73

The computed income turning point (ITP) measured in US$ is relatively high for each country compared with their current GDP per capita. Available statistics shows that only South Africa and Mauritius have reached this level of income among all the countries under study. Nigeria, Kenya and Seychelles are currently far below their computed income turning points. This implies that most Sub-Sahara African countries are still on the first half of the EKC. Hence, they are currently experiencing high and increasing environmental degradation.

6   Concluding Remarks and Recommendations

The myriad of environmental issues would mean selectively addressing, as we have done in this study, CO_2 which constitutes a major component of Green House Gas Emissions (GHGE). CO_2 is selected as a proxy to examine the behavior of pollutants at various levels of income.

From the perspective of Sub Saharan Africa, based on the empirical results from this study, economic development is intricately tied to environmental degradation. And at this point in time when Sub Sahara African countries are eagerly pursuing polices aimed at industrializing their economies; it would be unpopular to recommend a halt in the pursuit of economic development in order to save the environment. But it would be wise for government to mainstream the issue of the environment as part of the day to day policy formulation in order to fully address the balancing act between economic sustainability and environmental sustainability.

For this and other reasons, the countries in Sub-Saharan Africa are urged to pursue sensible economic development policies that reflect the pressing issues in environmental sustainability. It would be necessary for the countries to embrace a better understanding of the effects and problems of environmental degradation, particularly in relation to the anthropogenic (human activities) causes of global warming. In the process, provide necessary policy guidelines which reflect the social and economic exigencies in their societies and proffering necessary sustainable economic solutions.

The African leaders in the region are to undertake such mitigation efforts that are affordable and as long as they proactively implemented. By so doing, high levels of economic activities can still be sustained alongside a carefully and deliberately implemented set of environmental policies.

Other necessary policy engagement covering technology, education of the masses on environmental issues, capacity building and the building of infrastructures will go along way in achieving the balancing act between economic sustainability and environmental sustainability. Countries in the region should have a holistic view of the problem associated with environmental degradation in relation to economic development. That means integrating environmental concerns into economic development strategies across the various ministries or parastatals, such as the River Basins, Environment, Energy, Science and Technology, health, Education and Economic planning.

Since the majority of environmental models so far developed are for the benefit of the industrialized countries, African governments should develop their own African environmental data on atmospheric (green house) gases, climate change, and adaptable renewable energy technology. These governments should also undertake the establishment of meteorological infrastructure across the African continent to capture changes in climate and green house gas emissions, and build infrastructures that reduce green house gasses.

Finally, some of the policy recommendations will be better implemented if there are also collaborative efforts with other countries, international institutions and bodies. The Clean Development Mechanism (CDM) under the Kyoto Protocol allows African countries who are capable of achieving emission reduction targets to engage in sustainable development complemented with international resources. There are over 850 Clean Development Mechanism projects in developing countries, but only 23 are in Africa (AERC, 2008:10). This calls for a broadening of the scope since the more CDM credit earned, the more collaboration in the institution of eco-efficient projects in the affected countries.

References

AERC (2008): Climate change and economic development in Sub Sahara Africa. *Senior policy seminar Report AERC*, Signal press Kenya p. 10

Brock, William A. and M. Scott Taylor (2004): The green Solow model. *NBER working paper series No.* 10557.

Dasgupta, Susmite, Benoit Laplate, Hua Wang and David Wheeler (2002): confronting the Environmental Kuznets curve. *Journal of economic perspective 16(1) p. 147-168.*

Dijkfgraaf, Elbert and Herman R. J Vollebergh (2005): A note on testing for Environmental Kuznets curve With panel data. *Environmental and Resource economics 1(1)* p. 140-162.

Grossman, Gene M. and Elhanan Helpman (1997): *innovation and growth in the global economy*, Cambridge mass: MIT Press.

IBRD (1992): World development report 1992: *development and the environment* New York Oxford University press.

Kuznets, Simon (1995): Economic growth and income inequality. *American Economic Review 45(1)* p. 1-28.

Moomaw, William R. and Gregory C. Unrul (1997): Are Environmental Kuznets Curve misleading us? The Case of CO_2 emissions. *Environment and Development Economics 2(4)* p. 451-463

Panayotou Theodora (1993): Demystifying the Environmental Kuznets curve: turning a black box into a p Policy tool. *Environment and Development Economics 2(4)* p. 465-484

Roberts J. Timmons and Peter E. Grimes (1997): Carbon intensity and Economics Development 1962-91: A brief exploration of the Environmental Kuznets Curve. *World Development 25(2)* p. 191-198.

Selden Thomas M. and Oaqing Song (1994): Environmental quality and Development: is there a Kuznets Curve for air pollution Emission? *Journal of Environmental economics and Management 27(2)* p. 147-162.

Stern David I (2002): The rise and fall of the Environmental Kuznets curve. *World Development 32(8)* p. 1419-1439.

Torras Mariano and James K. Boyce (1998): Income, Inequality and pollution: A reassessment of the Environmental Kuznets curve. *Ecological Economics 25(2)* p. 147-160

Unruh Gregory C and William R Moomaw (1998): An alternative analysis of apparent EKC-Type Transition. *Ecological Economics 25(2)* p. 221-229

Vincent Jeffrey R. (1997): Testing for Environmental Kuznets curve within a developing country. *Environmental and Development Economics 2 (4)* 417-431

7.2: The Costs and Benefits of Environmental Protection: The Case of the Niger Delta Region of Nigeria

Milton A. Iyoha, Ph. D. (Yale)[1]

Abstract

The major environmental problems in the Niger Delta Region of Nigeria include air pollution, water pollution, flooding, soil erosion, agricultural land degradation, deforestation, fisheries depletion and habitat destruction, oil spills, biodiversity loss, and ecosystem degradation. Oil industry activities have been the main cause of environmental problems in this area. Expanded exploration and production activities by the oil companies have aggravated these problems that continue to impact negatively on the health of the people, their economic and social activities, sustainable development and ecological balance. Undoubtedly, the environmental problems in the Niger Delta have inevitably had a negative effect on the overall welfare, social and economic growth, and sustainable development of the citizens. In this paper, an attempt is made to analyze the benefits and costs of environmental protection in the Niger Delta Region of Nigeria by investigating the environmental effects of oil and gas industry activities on the Niger River Delta communities. These are found to be largely negative. The paper suggests that notwithstanding the contribution of the oil sector to the Nigerian economy, oil industry activities in the Niger Delta should be regulated to make them more environmentally friendly. Policy recommendations include: (i) the establishment and enforcement of strict environmental standards for air, water and land pollution; (ii) the encouragement of community participation and involvement in setting, monitoring and enforcing environmental standards and environmental protection; and (iii) adoption of measures to fight the extinction of terrestrial and aquatic species thus enhancing biodiversity for the benefit of current and future generations.

Key words: Ecosystem resilience, Environmental degradation, Environmental protection, Niger Delta Region of Nigeria

---

[1]   Professor of Economics, University of Benin, Benin City, Nigeria

1   Introduction

In recent years, global environmental problems of green house effect, acid rain, radioactive contamination, global warming, and sea-level rise have captured centre stage in international conferences and policy discussions. This is perhaps understandable as many of these problems, particularly, ozone layer depletion and global warming, threaten humanity with biological extinction. However, though perhaps less apocaplytic in consequences, environmental problems at the national and sub-national levels are also very important and inevitably have far-reaching consequences for sustainable growth and development. In Nigeria, these environmental issues include air and water pollution, desertification, agricultural land degradation, oil spills and blowouts, erosion and flooding, sea-level rise, fisheries depletion and habitat destruction, deforestation and forest degradation, biodiversity loss, and ecosystem degradation. Apart from desertification, which occurs in the northern part of the country, the other problems occur mainly in the southern parts of the country, especially in the Niger Delta region[i].

In Nigeria, in the Niger River Delta, oil and gas industry activities have continued to pose serious environmental problems affecting health, social and economic activities, sustainable development and ecological balance. The main environmental effects of oil and gas industry activities in the Niger Delta include:

i)   land degradation;
ii) air pollution;
iii) water pollution;
iv) deforestation and
v)   ecosystem degradation, resulting in biodiversity loss as a result of damage to terrestrial and aquatic life.

Though most of these effects are localized and most intense in the Niger Delta area, some, like air pollution, affect other parts of Nigeria and perhaps even some neighbouring countries. Similarly,

---

[i]   The Niger Delta region comprises nine states of the federation, viz., Abia, Akwa Ibom, Bayelsa, Cross River, Delta, Edo, Imo, Ondo, and Rivers. This region, which has an estimated population of 20 million is culturally and geographically diverse. Its cultural diversity is attested to by the fact that the Niger Delta comprises about 50 ethnic groups and over 3,500 communities who speak approximately 260 dialects. The major ethnic groups in the NDR are: Yoruba, Igbo, Ijaw, Urhobo, Isoko, Itsekiri, Ibibio, Annang, Kallabari, Ogbia, Egini, Ishan, Edo, Etsako, Ogoni, Idoni, Andoni, and Ekpaye. Geographically, the Niger Delta region covers about 70,000 square kilometres; over half of the region's surface is swampy, criss-crossed with creeks and dotted with small islands while the remaining portion is uplands, covered mainly by mangrove and rain forests. The Niger Delta region is the third largest wetland in the world and is covered by the world's third largest mangrove forest.

The Niger Delta region is the source of Nigeria's oil and gas resources, the pre-eminent earners of foreign exchange resources for the country. Yet, it is the least developed region of the country which is also threatened with ecological catastrophe because of the reckless exploitation of its oil and gas resources by foreign-owned oil companies.

mangrove swamp and rain forest destruction can have long-run consequences not only for the Niger Delta and other parts of the country but also for the entire world.

For the Nigerian economy as a whole, oil industry activities are a great benefit. In recent years, oil export revenues have accounted for approximately 97% of total exports and about 80% of total government revenues. (Iyoha, 2003). In many ways, oil has been the engine of economic growth in Nigeria since the early 1970s; rising oil prices have been associated with a booming economy while low world oil prices have precipitated a recession. A case in point was the early 1980s when the collapse of petroleum prices left Nigeria in a quagmire of economic problems—falling output, galloping inflation, rising fiscal deficits, increasing balance-of-payments deficits, and escalating external debt. Thus, we have a position where oil industry activities contribute positively to the growth of the Nigerian economy while at the same time they damage the environment and the ecology of the Niger Delta. Thus, the long-standing challenge has been how to develop strategies which would mitigate the environmental damage from oil and gas industry activities and alleviate the negative effects of oil exploitation on the Niger Delta region—thus promoting sustainable development in the region. For more, see Iyoha and Adamu (2002), Iyoha (2002), and Iyoha (2000).

In addition to this introductory section, the current paper contains five sections. Section II presents a discussion of the relationship between ecosystems, the environmental resource base and environmental degradation in order to throw light on the genesis, meaning and nature of environmental problems. Section III presents a quick overview of the benefits and costs of environmental protection in the Niger Delta Region of Nigeria. Section IV offers an examination of oil industry activities in the Niger Delta and analyzes the environmental consequences of these activities. It is also devoted to a theoretical analysis of the environment effects of oil industry activities in the Niger River Delta region. In particular, the Pigouvian framework of analysis is presented but modified to incorporate modern ideas that unidirectional externalities arise as much from market failure as from institutional (government and legal) failure. Section V then presents policy recommendations for mitigating the environmental degradation of the Niger Delta Region in the light of the preceding analysis. The last section is devoted to a summary of the paper and some concluding remarks.

2   Ecosystems, the Environmental Resource Base and Environmental Degradation

In order to fully understand the meaning, types, effects and consequences of environmental degradation, it is imperative to explain the concept of the environmental resource base and the role of ecological systems (ecosystems for short) which provide a wide range of services that are indispensable for all human activities and indeed essential for the survival of human and all other species. Ecological services, produced by ecosystems, are generated by interactions among organisms, populations of organisms, communities of populations, and the physical and chemical environment in which they reside. Dasgupta and Maler (1997) and Dasgupta, Folke and Maler (1994).

In particular, ecosystems are the sources of water, animal and plant food, and other renewable resources. Accordingly to Dasgupta *et al* (1994, p. 25).

> *They also maintain a genetic library, sustain the processes that preserve and regenerate soil, recycle nutrients, control floods, filter pollutants, assimilate waste, pollinate crops, operate the hydrological cycle, and maintain the gaseous composition of the atmosphere.*

The totality of all the ecosystems of the world represents a large part of what some authors call our "natural capital base" or more simply the *environmental resource base*. One may also simply call them environmental resources. Among these environmental resources we may list the earth's atmosphere; the sea; land; animal, bird, plant and fish populations; aquifers and other underground basins of water; and forests. From this partial list, it is obvious that the environmental resource base is of fundamental importance to man and needs to be carefully husbanded. Note that these environmental resources by and large are regenerative, i.e., they are renewable natural resources. They may be distinguished from non-renewable or exhaustible resources like minerals. However, though they are regenerative, their functioning can be damaged by misuse. Their self-regulatory mechanisms can be interfered with, leading to inefficient functioning, loss of function or even collapse. The earth's atmosphere is a case in point. Under normal conditions, the atmosphere's composition regenerates itself. However, the speed of regeneration depends on the rate at which pollutants are deposited in it and on the type and nature of the pollutants. Today, there is fear that CO_2 is being deposited in the atmosphere at too rapid a rate. Some fear that if this goes on, the ozone layer may collapse, an event which would let in the sun's lethal gamma rays, possibly leading to the elimination of all human life from the planet.

Environmental problems are therefore generally associated with resources that are regenerative (or renewable) but which are in danger of "exhaustion", "failure" or collapse from excessive use and/or degradation. This is as true of the atmosphere or sea or land as of forests and aquifers. In particular, note that degradation of the environmental resource base not only affects the quantity and quality of the services that are produced by ecosystems, it also threatens its resilience and carrying capacity. According to Dasgupta et al (1994, p. 26).

> *Resilience is the capacity of a system to recover from perturbations, shocks, and surprises. An ecosystem's reliance is its capacity to absorb disturbances without undergoing fundamental changes.*

Note that one of the central roles of biological diversity (often simply called biodiversity) is that it provides the ecosystem with resilience. Thus while we may want biodiversity for economic, social, ethical and esthetic reasons, one of its key functions is to provide the resilience, "integrity" or "stability" of ecosystems. According to Perrings et al (1992, p. 5).

> *The value of biological diversity thus lies in the value of the ecological services supported by the interaction between the organisms, population and communities of the natural environment, and the value of biodiversity loss reflects the sensitivity of ecological services to both the depletion and the deletion of species. There is a threshold of diversity below which most ecosystems cannot function.*

It is obvious that continued severe environmental degradation will eventually lead to collapse of the particular ecosystem and loss of its services. While low levels of environmental degradation may have no effect on the ecosystem, large amounts and levels will affect it. Unfortunately, the line between small and large damage to the environment is not known *a priori*. The truth is that the environmental resource base is a dynamic and complex living organism, since it consists of biological communities which interact with the physical and chemical environment in space and time. The interactions are often non-linear and subject to threshold effects, with discontinuities lying in wait in the flow of environmental services. Thus, the best strategy is to minimize environmental damage and thus safeguard our priceless and irreplaceable environmental resource base.

3   The Benefits and Costs of Environmental Protection in the Niger Delta: A Quick View

3.0 Preface: Countries strive to enhance and promote environmental protection because of its potential gains and benefits. However, there are also potential costs and dangers, particularly for developing countries that are poor and are attempting to accelerate their rate of development and industrialization. Logically, a rational decision can only be made on how far and rapidly to pursue the goal of environmental protection after weighing the benefits and costs. However, what should be borne in mind right from the beginning is that while the costs are short-term in nature and expected to be borne by the "polluters", the benefits are long-term and rather diffused (accruing to geographically dispersed individuals and communities in the entire Niger Delta region). Thus, in the long run, environmental protection is certain to be beneficial although the beneficiaries are geographically dispersed.

3.1   Benefits of Environmental Protection

The major benefits of environmental protection arise from the economic and social effects of reducing or eliminating environmental pollution in all its manifestations. In particular, mitigation of environmental degradation would reduce the loss of sustainable livelihoods in terms of destruction of farm lands and fish ponds, thus increasing the income and welfare of the indigenes. The greatest benefits no doubt arise from elimination of serious cases of environmental degradation which impact strongly on health and livelihoods. For example, reduction of oil spills will reduce water pollution and the destruction of fish ponds and farm lands. Another example is that reduction of gas flaring will improve the quality of the air and improve the health of the inhabitants. The main benefits of environmental protection therefore include:

(i)   elimination of the destruction of farm lands and fish ponds which are the principal sources of livelihood for farmers and fishermen in the Niger Delta Region;

(ii)   reduction of air pollution;

(iii)   elimination of acid rain;

(iv)   reduction of flooding and erosion;

(v)   improvement in the overall health and nutrition of the inhabitants;

(vi)   increase in productivity, quality of human capital and economic development; and

(vii)   increase in economic and social welfare of the inhabitants of the Niger Delta Region.

It is apparent that the benefits of environmental protection, which are dynamic and long-run in nature, will also be cumulative and should lead to increased investment, greater exploitation of scale economies, and enhanced growth over time. For further details, see Iyoha and Adamu (2002).

3.2   Costs of Environmental Protection

The costs of environmental protection are generally specific and short run in character. These costs are greater the more highly damaged the environment is before the adoption of remedial action and the more quickly one attempts to improve the environment. This is because the more pervasive the incidence of environmental degradation is, the more the distortion of prices and costs, the more the discrepancy between private and social costs and benefits, and the larger the total cost of remediation. The main costs of environmental protection may therefore be summarized thus:

(i)   total monetary cost of clean-up and other remedial action;

(ii)   loss of revenue and profit to private (oil) firms as a result of foregone output of oil and gas as a consequence of the need to reduce production for environmental reasons;

(iii)   loss of export revenue (foreign exchange) to the country as a result of reduction in oil and gas production;

(iv)   the total financial cost of reducing or eliminating gas flaring; and

(v)   the total financial cost of improving facilities (refineries, ships, etc.) and increasing the employment of skilled (specialist) workers in order to reduce the possibility of oil spills, blow-outs, etc.

4   Oil Industry Activities in the Niger Delta and Their Environmental Consequences

4.1   A Summary View

In considering the environmental effects of oil industry activities on the Niger Delta region and its environs, it is necessary to examine the activities of both oil companies and oil service companies. In addition, one must also analyze the environmental effects of petroleum refineries and petrochemical facilities. There are 3 refineries in the Niger Delta—1 in Warri and 2 in Port Harcourt. The old refinery in Port Harcourt is widely believed to be the source of much environmental damage consisting of:

i)   air pollution as a result of the release of major pollutants like sulfur oxides, nitrogen oxides, particulates, carbon monoxide, and hydrocarbons;

ii)   water pollution as a result of the release of hazardous sludge; and

iii)   contamination of surface water, soil, and groundwater as a result of spills of raw materials or leaks during processing.

Indeed, oil industry activities in the Niger Delta result in virtually all types of environmental degradation including especially:

i)   land resource degradation as a result of digging and dredging to clear vegetation at drilling sites and also from oil activity infrastructure;

ii)   renewable resource degradation. Under this heading, the most important issues are fishery stock depletion and habitat degradation; deforestation (both mangrove forest degradation and fresh-water forest degradation); and biodiversity loss arising from habitat destruction or degradation; and

iii) environmental pollution. Under this heading, the major issues of note are water contamination from both accidental oil spills and during the separation of oil and water at installations in the delta or at the coastal terminals; industrial pollution; release of toxic and hazardous substances, and release of untreated sewage. Also important is air pollution arising from gas flaring and other industrial activities including the operation of oil refineries. The major pollutants emitted from refineries include ammonia, sulfides, organic acids, chromium, and other metals.

Oil spills are of particular concern to communities in oil producing areas of the Niger Delta. Oil spills cause the mortality of bivalves such as oysters and fish. They have also contaminated and destroyed several hectares of mangrove swamp. Drilling fluids released from exploratory drilling are also an environmental hazard, as are drilling waste. Gas flaring has also raised a lot of concern in communities in the Niger Delta. Gas flaring is a major source of air pollution and releases methane gas to the atmosphere. The major consequences of gas flaring are

(i) the release of CO_2 and methane to the atmosphere;
(ii) acidification; and
(iii) water quality degradation.

Like other air pollutants, gas flaring poses serious health hazards to people living near the flaring sites. The ash residue also contributes both to soil and water degradation. Tragically, Nigeria has the dubious honour of being the foremost gas flaring country in the world.

In sum, oil industry activities in the Niger Delta have caused and continue to cause considerable ecological damage. In the main, oil industry activities have resulted in agricultural land degradation, air pollution, water pollution, and loss of terrestrial and aquatic biodiversity arising from habitat destruction. In addition to the negative economic, social and health consequence of environmental pollution on the inhabitants of the Niger Delta, oil industry activities pose a serious threat to the ecology of the wetlands and the long-term survival of their human and other species. For more details, see Iyoha (2002).

4.2 The Environmental Effects of Oil Industry Activities: A Theoretical Analysis

4.2.1 Externalities and Remedies

The point of departure of many economic analyses of environment effects of production activities is that pioneered by Pigou (see Pigou, 1920). Adopting the framework of a market economy and perfect competition, Pigou established that when there are externalities, private and social costs are not equated, nor are private and social benefits equal. Modern economic analysis supports these results but couches them in terms of the failure of perfect competition to achieve Pareto optionality in the presence of external effects in production or consumption, Henderson and Quandt (1971, p. 255). Environmental effects are clearly externalities since they are quintessential examples of circumstances where the costs and benefits are not fully reflected in potential or actual market exchanges. They arise as a result of incomplete or missing markets and are thus said to be caused by market failures. Helm and Pearce (1991, p. 2). This is also supported by Perrings, Folke and Maler (1992). According to them;

> *in a world of incomplete markets, environmental externalities will be pervasive. Externalities represent uncompensated costs or benefits of resource use that are born by someone other than the user. Perrings et al (1992, p. 13).*

Economic analysis, starting from Pigou's seminal work, demonstrates that the relevant marginal conditions for welfare maximization may be obtained by imposing appropriate taxes and subsidies. Pigouvian taxes and subsidies, possibly accompanied by lump-sum taxes and subsidies (to modify the distribution of income) are still widely recommended and used today.

However, modern analysis of environmental externalities has gone beyond the Pigouvian results. It is now widely believed that environmental externalities arise as much from market failure as from government failure. Government failure arises from the implementation of erroneous policies, ignorance, and incomplete information. A case often cited is the policy of the Brazilian government to give subsidies to firms for exploiting the Amazon rain forests. In retrospect, it is agreed that the policy was misdirected as it only served to accelerate deforestation and the destruction of the habitat of many species. The rapid reduction of rain forests is now widely believed to have serious negative global environmental consequences.

In general, as with market failure, government failure (sometimes called institutional failure) has resulted in an excessive use of environmental resources, leading to its degradation—with negative economic, social, health, and ecological consequences. Pigou's remedy was to use taxes and subsidies to modify imperfect markets and to alter costs, incentives and benefits. An extension of this approach is to create markets where they do not exist. This has led to the emphasis on the analysis and identification of property rights.

Note that government intervention is necessary to remove environmental disamenities and "bads", whether the genesis of the environmental externality is market failure or government failure. But there is still the question of the optimal method of intervention. Should the government use market instruments or should it use "command and control" laws and regulations? Most governments have chosen a pragmatic route of setting environmental standards, and then deciding how to attain them—either through pollution taxes, effluent charges, or even tradable pollution permits, or through bureaucratic regulation. Then there is the issue of who pays, the polluter or the victim? Analysis has shown that the optimal type of intervention will vary on a case-by-case basis and on the type of externality. However, since the bureaucratic solution of command and control requires a lot of information, it is likely to be more costly than a market-based solution which exploits the market's mechanisms for revealing information. But a presumption in favor of market-based policies does not mean that unfettered market forces should be given full rein. Rather, according to Helm and Pearce (1991, p. 22), it means that

> *. . . the market should be harnessed to generate the most efficient method of achieving desired pollution reductions. The role of the state is to regulate through command-and-control procedures in setting maximum pollution levels. The role of the market is to find the best method of achieving them.*

Market-based incentives or policy measures include taxes, charges, deposit-refund systems, tradable permits, and offset policies. The optimal measure to adopt will depend on the type of environmental problem and the type of economy, whether advanced or developing (as this will have implications for the level of development of markets). For more on these issues, see Ayres and Kneese (1969), Buchanan (1960), and Meade (1973).

4.2.2 The Welfare Optimum: Determination of the Optimal Amount of Environmental Protection

In what follows, as attempt will be made to derive the optimal amount of environmental protection or pollution abatement in a country or given locality. "Pollution" is used in a generic sense to include air pollution, water pollution or even land degradation. The optimal amount of environmental protection is defined as that which maximizes social welfare. More formally, it is that amount of environmental protection that maximizes the discounted sum of the net benefits from pollution abatement over a given horizon. Here, net benefits equal total benefits minus total costs, and the length of horizon may be infinite. The type of externality that is being dealt with is the "unidirectional" one, not the "reciprocal" kind. Reciprocal externalities are defined as those in which all parties who have rights of access to an environmental resource are able to impose costs on each others. An excellent example of reciprocal externalities is the much talked about "tragedy of the commons".

Unidirectional externalities are defined as those where the short-run external environmental costs or benefits of resource use are "one way", i.e., where a given party inflicts damage on others. In the Niger Delta, oil companies can and do pollute without compensating the people they harm. Indeed, we find many cases of unidirectional externalities such as oil spills (where the action of an oil company imposes costs on inhabitants of a community), air pollution from gas flaring, and water pollution (where effluent discharge etc. contaminates the surface water used by members of the community). Unidirectional externalities may arise from production or consumption activities. In the Niger Delta, they result mainly from the production activities of oil companies and oil service companies. Basically, oil industry activities impose costs on the community that are not fully reflected in actual market exchanges. For example, the costs of the pollution (of air, water and land) by oil companies are not fully internalized. The result is that the production of the "commodity" in question is excessive and so is the amount of pollution. If the oil companies are made to pay compensation for the pollution they cause (or otherwise fully bear the costs of environmental damage), equilibrium production will be less and the amount of pollution will fall.

In basic economic analysis, it is assumed that social welfare is maximized when the marginal social cost of producing a given commodity is equated to its marginal social benefit. However, in the presence of unidirectional externalities, private firms (like oil companies) will equate marginal private costs and benefits. It follows that the private optimum output, Q_p will exceed the socially optimum output, Q_s, since the costs of pollution are not internalized by the firm. Internalization of pollution costs, say by means of a Pigouvian tax, will shift the marginal cost curve upwards and reduce the equilibrium (private) production of the commodity. It can also be shown that the "optimum amount of pollution", that is, the pollution resulting from optimizing the social welfare function, will be greater than zero but less than what obtains without government intervention. For a demonstration of this result, see Iyoha (2002). This result, namely, that the "optimal amount of pollution" is not zero can be intuitively explained in several ways. One way of explaining it is that the cost of reducing pollution to zero will be astronomically high and hence not optimal from a social welfare point of view. Besides, small amounts of pollution are not likely to be too damaging as the ecosystem is primed to take care of such quantities. It is when pollution becomes excessive that it threatens the resilience of ecosystems and brings about the possibility of systemic collapse. Also, see Baumol and Oates (1988), Eskeland and Jinenez (1991), Maler (1974), and Seneca and Taussig (1984)

5   Mitigating Environmental Degradation in the Niger Delta Region of Nigeria: Some Policy Recommendations

5.1   Policy Framework and Policy Instruments

Going back to fundamentals, it could be said that there are two basic frameworks or approaches for mitigating the problem of environment degradation. They are:

(i) the market-based or decentralized decision-making framework and
(ii) the command-and-control or centralized framework.

The centralized or bureaucratic framework requires the getting of standards, monitoring them and having effective machinery for ensuring compliance. It would often entail the setting up of an elaborate bureaucracy and, in order for it to properly operate, it needs an efficient legal system and an army of dedicated inspectors. On the other hand, the market-based approach needs less information and leaner bureaucratic machinery. It is also less susceptible to rent-seeking and other acts of corruption.

Associated with these frameworks and approaches are policy instruments. According to Anderson (1996, p. 13),

> *The main policy instruments are familiar: environmental taxes on the main pollutants or sources of pollution; environmental laws and regulation, traditionally the instruments most favoured by governments; and for local pollution, negotiated arrangements, backed by local laws and institutions, between polluting and polluted parties.*

The use of markets to solve environmental problems has its origins in the Pigouvian insight that individuals and firms should be explicitly faced with the costs of environmental damage resulting from their activities. As much as possible, the polluter should be made to pay. Alternatively, we should make it worth his while to reduce pollution and other environmental damage. Indeed, increasingly, technologies are available which greatly reduce the amount of pollution released. Pollution abatement in electricity is a case in point, Anderson (1996). Note also that environmental taxes raise revenue; often they elicit widespread acceptance and compliance; and they allow the market to determine the most efficient methods of pollution production and control. Thus, once the "desired" level of pollution is set, the need for elaborate regulatory oversight is minimized. The fact that environmental taxes bring in revenue and improve government finances cannot be over-emphasized and may be an important source of revenue for local governments in the Niger Delta. Note that revenues from pollution taxes may be used to compensate those who suffer damage from the pollution or may be used by government, thus enabling it to scale back more distortionary forms of taxation. All these are achieved in addition to inducing lower emissions and better conservation of environmental resources.

However, since market and government failures vary on a case-by-case basis, so must the solution or remedy. This suggests a pragmatic blend of market approach and command-and-control regulations. Still, many analysts believe that the market approach should be encouraged whenever possible. According to Tietenberg (1991, p. 107).

> *While economic-incentive approaches to environmental control offer no panacea, they frequently do offer a practical way to achieve environmental goals more flexibly and at lower cost than more traditional regulatory approaches.*

Thus more attention should be given to exploring the usage of market-based incentives like taxes, charges, deposit-refund systems, and tradable permits. It is noteworthy that pollution charges have been successfully applied in many developing countries such as Brazil, Colombia, Jamaica, Mexico, Chile, Peru, China, and Malaysia. See Steer (1996) and Thomas and Belt (1997).

Finally, for local environmental problems as in the Niger Delta, negotiated arrangements between polluting and polluted parties can be quite effective in reducing environmental damage. These types of agreements, backed by local laws and institutions, have been successfully used to mitigate environmental degradation in Japan, Anderson (1996), Thomas and Belt (1997). It is not surprising that this strategy is now increasingly canvassed given the current emphasis on local empowerment, community involvement and grassroots participation as a means of ensuring the sustainability of policies and projects.

5.2 Policy Recommendations:

Environmental pollution causes considerable health damage to inhabitants of the Niger Delta. Land and water pollution have also had negative impacts on agriculture and fishing—thus further impoverishing the inhabitants of the Niger Delta. Continued loss of terrestrial and aquatic biodiversity due to habitat destruction is a ticking ecological time bomb not only for Niger Delta residents but for all Nigerians. For more, see Egbon (1996), Pearce and Warford (1993), and World Bank (1995). Therefore, notwithstanding the contribution of the oil sector to the Nigerian economy, oil industry activities in the Niger Delta must be regulated to make them more environmentally friendly. In order to mitigate the environmental degradation of the Niger Delta, the following policies and strategies are recommended for adoption and implementation:

1. Strict environmental standards for air, land, and water pollution should be enacted and enforced. The Environmental Protection Agency should be strengthened for this task. See Federal Government of Nigeria (1988).
2. As much as possible market based instruments like pollution taxes and effluent charges should be utilized. This will economize on the use of bureaucracy and reduce the cost (human and material) of enforcement. Also revenue obtained from pollution taxes should be used for environmentally benign projects or to compensate inhabitants of the Niger Delta Region who have suffered environmental damage.
3. An attempt should be made to mainstream environmental concerns in national economic policies. This will promote visibility and sustainability of environmental policies.
4. An attempt should be made to encourage community participation and involvement in setting, monitoring and enforcing environmental standards. One way to achieve this is by negotiated

agreements backed by local laws and institutions between the polluting party (oil companies) and the affected communities in the Niger Delta.

6   Summary and Conclusion

In this paper, an attempt has been made to analyze the benefits and costs of environmental protection in the Niger Delta Region of Nigeria. This was accomplished by first investigating the environmental effects of oil and gas industry activities on the Niger River Delta communities. Oil industry exploration and production activities were found to inflict serious ecological damage on the Niger Delta. The main environmental effects include agricultural land degradation, mangrove and rain forest degradation, air pollution, water pollution, and habitat destruction leading to loss of terrestrial and aquatic biodiversity. Continued environmental degradation causes considerable health damage while continued biodiversity loss threatens ecosystem integrity, which in turn threatens the Niger Delta with ecological catastrophe. Meanwhile the negative economic and social effects of the environmental damage have led to a volatile political situation in the Niger Delta. Thus, the sooner the environmental dis-amenities of oil industry activities are tackled and the damage mitigated, the better for everyone.

As a prelude to suggesting policies for mitigating the environmental degradation of the Niger Delta, this paper presents a theoretical analysis of the environmental effects of oil industry activities in the region. The point of departure is the theory of externalities. Indeed, environmental effects are best viewed as externalities, i.e, cases where the costs and benefits of production or consumption activities are not fully reflected in potential or actual market exchanges. In the Niger Delta, we have a case where oil industry activities impose a cost or damage on the community which is not reflected in the market price of the commodity. In technical terms, what occurs in the Niger Delta is a case of unidirectional production externality. One way of remedying the damage, first suggested by Professor Pigou, is a tax on the polluter. The tax results in the internalization of the cost, reducing the equilibrium output and reducing the environmental pollution. Given the right amount of tax (or subsidy), the "optimal" amount of pollution can be obtained. In general, this pollution level (i.e., the socially "optimum amount of pollution") will be greater than zero but less than the free market level of pollution. The tax revenue obtained may be used to compensate those damaged by the environmental pollution or utilized to finance environmentally benign projects.

The paper ends by proposing strategies, policies and measures to mitigate the environmental damage inflicted on the Niger Delta by oil industry activities. They include the following:

(i)   setting of strict environmental standards and their enforcement through the Environmental Protection Agency or by state and local laws;

(ii)   mainstreaming environmental concerns;

(iii) applying market-based instruments like pollution taxes and effluent charges to induce pollution abatement; and

(iv)   involving the community in negotiated arrangements backed by local laws and institutions, between polluting oil companies and affected inhabitants. The truth of the matter is that the technology for significant *pollution reduction* already exists. All that is left is to use the "carrot and stick" approach to make oil companies use them and thereby improve the environment for present and future generations.

REFERENCES

Anderson, D. 1996. "Energy and the Environment: Technical and Economic Possibilities", *Finance and Development*, Vol. 33, No. 2 (June).

Ayres, R.V and A. Kneese. 1969 "Production, Consumption and Externalities", *American Economic Review* Vol 59, No. 3.

Baumol, W and W. Oates 1988 *The Theory of Environmental Policy* (Cambridge: Cambridge University Press).

Buchanan, J.M 1960. "External Diseconomies, Corrective Taxes and Market Structure", *American Economic Review* Vol. 59, No. 1 (March).

Dasgupta, P. and K.G. Maler. 1997. "The resource basis of production and consumption: An economic analysis". In Dasgupta, P. and K.G. Maler (Eds.), *The environment and emerging development issues*. New York: Oxford University Press.

Dasgupta, P., C. Folke, and K.G. Maler. 1994. "The environmental resource base and human welfare". In Lindahl-Kiessling, K. and H. Landberg (Eds.), *Population, economic development, and the environment*. New York: Oxford University Press.

Egbon, P.C. 1996 "Environmental Policy Analysis: The Case of Nigeria", In Egbon, P.C. and B. Morvaridi (eds) *Environmental Policy Planning*. Ibadan: NCEMA.

Eskeland, G.S and E. Jinenez 1991 "Curbing Pollution in Developing Countries" *Finance and Development*, Vol 28, No. 1. (March).

Federal Government of Nigeria. 1988 Federal Environmental Protection Agency Decree (58). Lagos Government Printers.

Helm, D and D. Pearce 1991 "Economic Policy towards the Environment: An Overview", In Helm, D. (ed)., *Economic Policy Towards the Environment* (Oxford: Blackwell Publishers).

Henderson, J. M. and R.E Quandt. 1971. *Microeconomic Theory: A Mathematical Approach* (New York: McGraw-Hill)

Iyoha, M.A. 2003. "An overview of leading issues in the structure and development of the Nigerian economy since 1960", in *Nigerian Economy: Structure, Growth and Development*, M.A. Iyoha and C.O. Itsede (Eds.). Benin City: Mindex Publishing.

Iyoha, M. A. (2002). "The environmental effects of oil industry activities on the Nigerian economy: A theoretical analysis". In *The Petroleum Industry, the Economy and the Niger Delta Environment*, edited by C.O. Orubu, D.O. Ogisi and R.N. Okoh. Proceedings of a National Conference on the Nigerian Petroleum Industry and the Niger Delta Environment.

Iyoha, M.A. 2000. "The environmental effects of oil industry activities on the Nigerian economy: A theoretical analysis". Processed.

Iyoha, M. A. and P. A. Adamu (2002). "A theoretical analysis of the effect of environmental problems on economic development: The case of Nigeria". *Nigerian Economic and Financial Review,* vol. 7, no. 1. June.

Maler, K.G 1974 *Environmental Economics: A Theoretical Inquiry.* (Baltinore: Johns Hopkins Press).

Meade, J.E 1973. *The Theory of Externalities* (Geneva).

Pearce, D.W. and J.J. Warford. 1993. *World without end: Economics, environment, and sustainable development.* Washington, DC: The World Bank.

Perrings, C., C. Folke, and K.G Maler 1992. "The Ecology and Economics of Biological Diversity: Elements of a Research Agenda". Beijer Discussion Paper Series No. 1, Beijer International Institute of Ecological Economics.

Pigou, A.C. 1920. *The Economics of Welfare* (London: Macmillam).

Seneca, J.J. and M.K. Tausig 1984 *Environmental Economics* (Englewood Cliffs, N.J: Prentice-Hall).

Steer, A. 1996 "Ten Principles of the New Environmentalism", *Finance and Development* Vol. 33, Number 4, (December).

Thomas, V and T. Belt. 1997. "Growth and the Environment: Allied or Foes', *Finance and Development* Vol. 34, No. 2 (June).

Tietenberg, T.H 1991 "Economic Instruments for Environmental Regulation". In Helm, Dr. (ed.) *Economic Policy Towards the Environment* (Oxford: Blackwell Publishers).

World Bank. 1995. *Defining an environmental development strategy for the Niger Delta,* vols. 1 & 2. Washington, DC: The World Bank.

7.3 Advancing Guidelines for Environmental Safety through Ecotoxicological Research

[1]Lawrwnce I.N. Ezemonye and [2]Tongo Isioma

Email: ezemslaw@yahoo.com, isquared27@yahoo.com

Abstract

Ecotoxicological research is a prospective and retrospective tool which predicts environmental risk of contaminants and also evaluates environmental damage; it is a requirement for protecting the environment and setting of environmental standards. Predictive ecotoxicology emphasizes the probable environmental outcome of exposure to toxins, rather than the mere appraisal of existing damage and in so doing raises some complex but interesting ethical issues. Studies on ecotoxicology provide screening tools to identify chemical concentrations in environmental media that could be of potential risks to ecological receptors. Ecotoxicology has focused almost exclusively on countries and ecosystems in temperate zones. Tropical ecosystems, which contain as much as 75% of the global biodiversity, have been neglected. Tropical ecosystems are under increasing threat of development and habitat degradation from population growth and urbanization, agricultural expansion, industrialization, and technological development. Some of these activities also lead to the release of toxic substances into the environment. Little research in ecotoxicology has been carried out in tropical environments. In Nigeria, only a few chemicals have been ecologically tested for safety in spite of their ecological and environmental impact. This paper presents an overview of some of the ecotoxicological studies carried out in Nigeria with the aim of providing a battery of biomarkers and estimated safe limits for some industrial chemicals and pesticides in the aquatic ecological regime in Nigeria. It also advances guidelines for regulatory surveillance and monitoring of the aquatic ecosystem in Nigeria.

Keywords: Ecotoxicology, Hazardous chemicals, Environmental Safety, Safe limits

[1]   National Centre for Energy and Environment (NCEE), Energy Commission of Nigeria,University of Benin
[2]   Department of Animal and Environmental Biology (AEB) University of Benin

Introduction

There are several major environmental issues in Nigeria amongst which is the release of toxicants by different anthropogenic activities. Contamination of the environment with a wide range of pollutants has become a matter of concern over the last few decades (Ezemonye *et al.*, 2007, Ezemonye and Tongo, 2009). Industrial chemicals and pesticides are released indiscriminately through usage, spillage and during transportation. The environment may therefore be extensively contaminated with chemicals released from industrial and anthropogenic activities (Velez and Montoro, 1998; Conacher, *et al.*, 1993). The rate of intensity of man's impact on the environment is so great that a collective responsibility for its protection is imperative. (Cairns 1980).

Following the Koko saga of 1987, where there was illegal dumping of toxic wastes in Koko, in the former Bendel State, the Nigerian government through a series of evolutions established various regulatory bodies to ensure compliance of environmental safety guidelines. These bodies include the Federal Environmental Protection Agency (FEPA) established in 1988 and charged with the overall responsibility of protecting and developing the Nigerian environment. The Department of Petroleum Resources (DPR), an arm of the Ministry of Petroleum Resources responsible for setting out comprehensive standards and guidelines to regulate the execution of oil and gas projects with proper GREEN considerations for the environment. The Federal Ministry of Environment (FMENV) and their states counterpart and recently in 2007, the establishment of the National Environmental Standards and Regulations Enforcement Agency (NESREA), responsible for the protection and development of the environment, biodiversity conservation and the sustainable development of Nigerian's natural resources. National Oil Spill Detection and Response Agency (NOSDRA) for the detection and cleaning up of oil spill.

These regulatory bodies use safety guidelines for the protection of the environment. Existing guidelines in Nigeria include: EIA Sectorial Guidelines, National Effluent Limitation Regulations 1991, Pollution Abatement in Industries and Facilities Generating Wastes Regulations of 1991 and the Solid and Hazardous Wastes Management Regulation of 1991. These guidelines were usually prepared using site-specific scientific studies (SSSS) and Ecotoxicological considerations, which encompass both prospective and retrospective tools for the perdition of environmental risk of contaminants and evaluation of ecological damage (DPR, 2002).

Ecotoxicology

Ecotoxicology is the science of predicting effects of potentially toxic agents on natural ecosystems and on nontarget species (Hoffman *et al.*, 2003). It is a logical extension of the field of toxicology which is the science of the effects of poisons on individual organisms, to the ecological effects of chemical of pollutants (Hoffman, 2003; Cairns,2003. Literally, ecotoxicology applies to all biological organization levels, from a single species embedded in its niche to the biosphere, including humans. The use of the prefix 'eco' implies that tests will use endpoints characteristic of levels of biological organization higher than single species (Hoffman, 2003).It is a subject involves the direct study of toxic effects of xenobiotic agents organisms. The primary purpose of ecotoxicology is to provide a means of predicting the probability of harm from the use of chemicals or other environmental stressors (*e.g.*, heat or suspended solids) upon complex natural systems. In a more restrictive but useful sense, ecotoxicology assesses the effects of toxic substances on ecosystems with the goal of protecting entire ecosystems, and not merely isolated components (Hoffman *et al.*,2003).

It is a requirement for protecting the environment for effect of pollutants. Predictive ecotoxicology emphasizes the probable environmental outcome of exposure to toxins, rather than the mere appraisal of existing damage. Ecotoxicologial considerations provide screening tools to identify chemical concentrations in environmental matrices that could be of potential risks to ecological receptors. It helps professionals/regulators involved in environmental protection to determine site-specific safe concentrations of chemicals used and thus provide for their disposal at concentrations not detrimental to aquatic and terrestrial life. The prediction of environmental risk of contaminants (prospective) and the assessment of environmental damage (retrospective) is an important goal of ecotoxicology in protecting the form and function of ecosystems through the setting of environmental standards.

Unfortunately, Ecotoxicological trials are usually not used as environmental safety tools in establishing guidelines in developing countries rather it has been exclusively in temperate ecological zones. Tropical ecosystems, which face threats of habitat degradation from population growth, urbanization, agricultural expansion, industrialisation and technological development, have been completely neglected in this region. Ecotoxicology considerations in Nigeria are still in its infancy and needs to be greatly explored. The central theme of this presentation is a clarion call: (1) to protect natural habitats from alterations due to the effects of toxic substances (2) to develop standards and guidelines from ecological data from researches done locally in Nigeria (3) to build up a national information bank on ecotoxicological studies in Nigeriacollating research efforts from local researchers in Nigeria and finally (4) to mandatory involve Ecotoxicology as one of the key environmental safety tool for development of environmental standards and guidlines.

Approaches/Strategies for Ecotoxicological Considerations

Ecotoxicologists are interested in assessing the health of ecosystems. Monitoring of the ecosystem health requires the use of different strategies amongst which are ecotoxicological trials and Ecologial Risk Assessment. Assessing ecological health of species, populations, communities and ecosystems using such trials involves evaluating the current status and changes over time making ecotoxicology a retrospective and prospective tool in Ecological Risk Assessment (Bartell, 2006).

Ecotoxicological toxicity tests can be used evaluate the effects of contamination on the survival, growth, reproduction, behoviour or other attributes of the of the organism. Toxicity tests measure lethal and sublethal effects and these effects are known as measurement endpoints. Generally, toxicity tests are divided into two broad categories:

1. **Acute toxicity tests**; are short term tests that measure the effects of exposure to relatively high concentration of chemicals. The measurement endpoint generally reflects the extent of lethality (Baker,1989).

2. **Chronic toxicity tests**; are generally long-term test that are designed to measure the effects of toxicants to organisms over a significant portion of the organism's life cycle, typically one tenth or more of the organism's lifetime. Chronic studies evaluate the sublethal effects

of toxicants on reproduction, growth, and behavior due to physiological and biochemical disruptions (Hoffman,2003).

Biomarkers

Ecotoxicological trials involve the use of a battery of biomarkers to provide early warnings (Ricketts *et al.,* 2003). Concisely, biomarkers can be defined as the biological (physiological, histological, cellular or biochemical) manifestations of pollutant stress (Huggett *et al.,* 2002). Firstly, biomarkers are, responsive only to the biologically active fraction of accumulated body burden of one or more toxicants and characterize bioavailable fractions of environmental chemicals. Secondly, biomarkers integrate the interactive effect of complex mixtures of chemicals experienced by organisms in ecosystems impacted by modern industrial and agricultural chemicals. (Bartell, 2006). Biomarkers can be measured at molecular, biochemical, cellular or physiological level of biological organization (Ricketts *et al.,* 2003).

During the past 20 years, substantial efforts have been made to develop in applying biomarkers for use in Ecotoxicology and Ecological Risk Assessment. The current trend in the last few years has been geared towards developing cheaper and faster tests using biomarkers (Clements, 2000). This has resulted in the development of biomarkers for early warning indicator responses before noticeable effects on individual performance and population dynamics are observed. Biomarker assessment are recommended alternatives to whole animal teats, that supports the 3Rs concept of reducing, refining and replacing whole animal tests. Biomarker responses are rapid (hours to days) and can be determined in situ yielding integrative responses under actual field conditions. Biomarkers offer one possible solution to the recognized limitations of extrapolating the results of single-chemical, single-species laboratory toxicity assays in assessing ecological risk (Bartell, 2006).

A variety of markers have been developed for determining the toxicological effects of pollutants released into the environment and the biological quality of studied sites (Table 1). Individual responses such as survival, reproduction and locomotion are routinely used as endpoints but they do not give early warming information on the impacts of the contaminants and are not easily adapted to monitoring in field conditions, they also have low sensitivity and are unspecific (Booth *et al.,* 1998).

Table 1: Biomarkers that have been used to asses impairments of biological function in organisms.

Biomarker	Tissue	Use	Reference
Mixed-function oxidases	Liver	Indicator of exposure to organic chemicals such as PAH and PCBs	Hyne and Marker, 2003
Gluthathione S-transferase	Liver	Indicator of exposure to pesticides and metalloids	Tongo,2010 Ikpesu, 2010 Yang,1996, Ezemonye and Tongo,2010,
Acetylcholinesterase	Brain	Indicator of exposure to organophosphates and carbamate pesticides	Tongo,2010 Ikpesu, 2010
Stress proteins	Various	Indicator of cells experiencing stress	Tongo,2010 Ikpesu, 2010
Glycogen level	Various	Indicator of exposure of oxidative stress	Ezemonye and Ilechie,2007 Ezemonye and Tongo, 2009
Metalothionein	Various	Indicator of exposure to metals	Hamer et al., 1986

Ecotoxicological trials in Nigeria

Ecotoxicological trials in Nigeria have not been given serious consideration compared to research initiatives in the parent fields of Experimental Ecology and Toxicology (Ezemonye and Tongo, 2010). Only a few chemicals have been ecologically tested for safe limits in spite of their ecological and environment impact. Worse still, the adoption of standards and guidelines obtained from data from other countries has proven to be ineffective as a result of differences in environmental conditions and species assemblage.

The current issuance of safe permit for the use of chemicals in Nigeria base on their Material Safety and Data Sheets (MSDS) which does not contain any ecotoxicological consideration has not helped matters.

This paper presents some ecotoxicological trials on commonly used chemicals in Nigeria. These chemicals range from insecticides, herbicides, Industrial dispersant and demulsifies that are used on a daily bases for industrial and agricultural activities. Table 2 shows estimated safe concentrations, relevant biomarkers and toxicity profiles for commonly used chemicals in Nigeria.

TABLE 1: ECOTOXICOLOGICAL PROFILE OF COMMONLY USED CHEMICALS IN NIGERIA, WITH THE VARIOUS INDICATOR SPECIES, BIOMARKERS AND SAFE LEVELS.

A.   NEATEX CR 486   Application: INDUSTRIAL DETERGENT

	Organism Tested/ Bioindicator	Species	Contaminant Concentration	Life Stage In days	Duration of Exposure	LC$_{50}$	Estimated Safe Conc.	Endpoint/ Biomarker	Effect	Comment	Reference
1	**FISH**	*Tilapia guineensis*	6.25, 12.5, 25, 50, 100 mg/l	7,14,28	96 Hours	8.79-82.42 mg/l (Fresh Water) 15.42-46.52 mg/l (Brackish water)	0.55-2.02 mg/l Fresh Water) 0.74-2.45 mg/l(Brackish water)	Mortality	Estimated 96 Hours LC$_{50}$ was dependent on species age, exposure duration and environment	High mortality of 7-day old test organism to the chemical provides a rational for monitoring of the waters of the Niger Delta	Ezemonye *et al*, 2007b
2	**SHRIMPS**	*Desmoscais trisinosa*	31.35, 62.5, 125, 250, 500 mg/kg		10 days	139.49 mg/kg (Fresh Water)	13.95 mg/kg	Mortality Mortality	Observed % mean mortality of the test organism to the chemical was significantly different from the control suggesting that mortality may be induced by the effect of the chemical. LC50 varied with species type.	Values from the study is an indication that the chemicals have the potential to cause acute lethal toxicity	Ezemonye *et al*, 2007a
		Palaemonetes africanus	1.35, 62.5, 125, 250, 500 mg/kg		10 days	259 mg/kg (Brackish water)	5.9 mg/kg		"	"	Ezemonye *et al*, 2007a
3	**Earthworm**	*Aporrectodea longa*	62.5, 125, 250, 500 mg/kg1	14	96 Hours	511.32 mg/kg	51.13 mg/l	Mortality	Percentage motality increased with increase concentration and exposure duration.	Values from the study is an indication that the chemicals have the potential to cause acute lethal toxicity	Ezemonye *et al*, 2006

INDUSTRIAL CHEMICALS

B. NORUST CR 486   **Application: CORROSION INHIBITOR**

	Organism Tested/ Bio-indicator	Species	Contaminant Concentration	Life Stage In days	Duration of Exposure	LC_{50}	Estimated Safe Conc.	Endpoint/ Biomarker	Effect	Comment	Reference
1	**FISH**	*Tilapia guineensis*	6.25, 12.5, 25, 50, 100 mg/l	7,14,28	96 Hours	5.55-20.21 mg/l (Fresh Water) / 7.35-24.50 mg/l (Brackish water)	0.55-2.02 mg/l / 0.74-2.45 mg/l	Mortality	Estimated 96 Hours LC50 was dependent on species age, exposure duration and environment	High mortality of 7-day old test organism to the chemical provides a rational for monitoring of the waters of the Niger Delta	Ezemonye *et al*, 2007b
2	SHRIMPS	*Desmoscais trisinosa*	31.35, 62.5, 125, 250, 500 mg/kg		10 days	78.61 mg/kg (Fresh Water)	7.86 mg/kg	Mortality	Observed % mean mortality of the test organism to the chemical was significantly different from the control suggesting that mortality may be induced by the effect of the chemical. LC50 varied with species type.	Values from the study is an indication that the chemicals have the potential to cause acute lethal toxicity	Ezemonye *et al*, 2007a
		Palaemonetes africanus	31.35, 62.5, 125, 250, 500 mg/kg		10 days	136.53 mg/kg (Brackish water)	13.65 mg/kg	Mortality	"	"	Ezemonye *et al*, 2007a
3	**Earthworm**	*Aporrectodea longa*	62.5, 125, 250, 500 mg/kg1	14	96 Hours	207.61 mg/kg	20.76 mg/l	Mortality	Percentage mortality increased with increase concentration and exposure duration.	Values from the study is an indication that the chemicals have the potential to cause acute lethal toxicity	Ezemonye *et al*, 2006

C. LAUNDRY EFFLUENT

	Organism Tested/ Bio-indicator	Species	Contaminant Concentration	Duration of Exposure	Life Stage In days	LC_{50}	Estimated Safe Conc.	Endpoint/ Biomarker	Effect	Comment	Reference
1	**FISH**	*Heteroclarias species*	5%, 6%, 7%, 8%, 9%, 10%	96 Hrs	Fingerlings	9.7%	0.97%	Mortality	Percentage mortality increased with increase concentration and exposure duration.	Values from the study is an indication that the chemicals have the potential to cause acute lethal toxicity	Ezemonye *et al*, 2003.
2	**INSECT**	*Chironomus travalensis*	5%, 6%, 7%, 8%, 9%, 10%	96 Hrs	Larvae	10%	1%	Mortality	"	"	Ezemonye *et al*, 2003.
3	**AMPHIB-IAN**	*Bufo regularis*	5%, 6%, 7%, 8%, 9%, 10%	96 Hrs	Tadpole	8.2%	0.82%	Mortality	"	"	Ezemonye *et al*, 2003.

D.	SEWAGE EFFLUENT										
1	FISH	*Tiliapia Zilli*	"	"	Fingerlings	84%	0.84%	Mortality	Percentage mortality increased with increase concentration and exposure duration.	Values from the study is an indication that the chemicals have the potential to cause acute lethal toxicity	Ezemonye and Olomu-koro, 2003.
		Epiplatys sexfasciatus	"	"	Fingerlings	100%	10%	"	"	"	"
2	SHRIMP	*Caridina Africana*	0.5, 5, 10, 20, 50, 80, 100%	96 Hrs	-	31%	0.31%	"	"	"	Ezemonye and Olomu-koro, 2003.

E.	PORTLAND CEMENT										
1	FISH	*Oreochromis niloticus*	19.60, 78.20, 156.80, 312.5 (mg/l)	96 Hours	14 days	41.21 mg/l	0.412 mg/l	Mortality	Percentage mortality increased with increase Portland cement concentration. Varying behavioural patterns observed in the fish include: erratic swimming, loss of reflexes, loss of equilibrium, paleness of skin and gasping for air.	The most common gill changes of Portland cement powder in solution were destruction of gill lamella, epithelial hyperplasia and epithelial hypertrophy.	Adamu, et al., 2008

F.	POTASSIUM PERMANGANATE										
1	FISH	African catfish *Clarias gariepinus*	19.60, 39.10, 156.80, 312.5 mg/l	96 Hours	14 days	3.02 mg/l	0.302 mg/l	Mortality Avoidance Response	Insensitivity to stimuli of touch and loss of stability were observed in test fish	Fingerlings were used for the experimental protocol	Kori-Siak-pere, O., 2008

G.	CRUDE OIL									
	Organism Tested/ Bioindicator	Species	Contaminant Conc.	Life Stage	LC50	Safe Conc.	Endpoint/ Biomarker	Comments	Reference	
	MOLLUSCS	*Carasostrea gasar*	0.1ppm, 1ppm, 10ppm 100ppm, 500ppm, 1000ppm	Larvae	135ppm	13.5ppm	Mortality	Small Oysters are more resistant than large Oysters to oil pollution	Daka and Ekweo-zor, 2004	

H.	DIESELS FUEL									
	MAMMAL	*Rattus rattus*	65g, 87g, 109g 131g	Adult	70.6g	7.06g	Mortality, histopathological alterations	Histopathaological alterations of lungs showed induced lesions	Dede and Kagbo, 2001	

I. GAMMALIN 20									
MAMMAL	Guinea Pig	0.82g/kg, 1.63g/kg, 3.25g/kg, 6.5g/kg	Adult	1.87g/kg	0.187g/kg	Death, histopathological changes	There were indications that Lindane is tumor promoter		Dede and Dogara, 2004
J. CHLOROPYRIFOS									
FISH	*Tilapia guineensis*	0.1mg/l, 0.5mg/l 0.025mg/l, 0.0125mg/l	Fingerlings	0.002mg/l	0.0002mg/l	Mortality			Chindah *et al*, 2004
K. AGROLYSER									
FISH	*Heterobranchus bidorsalis,* *Clarias gariepinus*	1400mg/l, 1600mg/l, 1800mg/l, 2000mg/l 2200mg/l	Fingerlings	1688.58mg/l	168.858mg/l	Mortality	The 96hrs LC50 is much lower than the recommended, hence if this level is maintained in the field, the possibility of toxic effects may be rare		Gabriel *et al*, 2009
L. PARAQUAT									
RAT	*Rattus norvegicus*	0.09g/kg, 0.18g/kg, 0.35g/kg, 0.70g/kg, 1.00g/kg	Adult	0.45g/kg	0.045g/kg	Mortality	There is need for caution in the use of paraquat in occupational, recreational industrial areas as accidental exposure to man or animal could cause possible toxic effect.		Dede *et al*, 2007

AGRICULTURAL PESTICIDES

| A. | ATRAZINE | | Application: Herbicide | | | | | | | FEPA standard for water and land: < 0.01 | |

	Organism Tested/ Bio-indicator	Species	Contaminant Concentration	Life Stage	Duration	LC$_{50}$	Safe Conc.	Endpoint/ Biomarker	Effect	Comment	Reference
1	AMPHIBIAN	*Ptychadena bibroni* (Frog)	200,400,600 (µg/l)	14 days	96 Hours	230.62 µg/l	23.01 µg/l	Mortality Avoidance Response	Substantial mortality and incidence of abnormal avoidance response occurred at higher concentration	High mortality of 14 day old test organism to Atrazine provides a rational for monitoring of the waters of the Niger Delta	Ezemonye and Ilechie, 2006
		"	"	21 days	"	305.9µg/l	30.59 µg/l	"	"	"	Ezemonye and Tongo, 2009
		"	"	28 days	"	431.3µg/l	43.13 µg/l	"	"	"	Ezemonye and Tongo, 2009
		"	3,30, 100	7,14,21,28 days	28 days			Glycogen level	Depletion in the glycogen levels in organisms exposed to chemicals compared to the control	Depletion in the glycogen levels indication of probable toxicological effect as observed in oxidative stress.	Ezemonye and Tongo, 2009

| B. | BASUDIN | | Application: Insecticide | | | | | | | | |

	Organism Tested/ Bio-indicator	Species	Contaminant Concentration	Life Stage In days	Duration of Exposure	LC$_{50}$	Estimated Safe Conc.	Endpoint/ Biomarker	Effect	Comment	Reference
1	AMPHIBIAN	*Ptychadena bibroni* (Frog)	0.1, 1, 10, 25 µg/l	14	96 Hours	0.860 µg/l	0.086 µg/l	Mortality Avoidance Response	Substantial mortality and incidence of abnormal avoidance response occurred at higher concentration	High mortality of 14 day old test organism to Basudin provides a rational for monitoring of the waters of the Niger Delta	Ezemonye and Ilechie, 2006

C. DIAZINON   **Application: Insecticide**

1	AMPHIB-IAN	*Bufo regularis* (Toad)	0.25, 0.50, 0.75 and 1 mg/l.	Adult	96 Hours	0.44 mg/l)	0.04 mg/l	Mortality Bioconcentration Behavioural and morphological changes	Substantial mortality and incidence of Behavioural and morphological alterations occurring at higher concentrations	significant positive correlation between tissue concentration and mortality with dose-dependent deformities and behavioural abnormalities. More pronounced poisoning symptoms at higher concentrations.	Ezemonye and Tongo, 2010a
		Bufo regularis (Toad)	0.01, 0.02, 0.03 and 0.04 µg/l	Adult	28 Days			Glutathione-S-trsnsferase (GST)	Alteration in Glutathione-S-trsnsferase (GST) activity in liver, brain serum, lungs, GIT tissues	GST is an efficient biomarker for Diazion toxicity	Ezemonye and Tongo, 2010b
2	FISH	*Clarias gariepinus*	3.00, 3.25, 3.50, 3.75 and 4.00 (mg/l)	Fingerlings	96 Hours	3.12 mg/l	0.31mg/l	Mortality	Percentage motality increased with increase in Diazinon concentration.	-	Iyang R. Ph.D Thesis 2009. RSUST.

D. ENDOSULFAN   **Application: Insecticide**

	AMPHIB-IAN	*Bufo regularis* (Toad)	0.25, 0.50, 0.75 and 1 mg/l.	Adult	96 Hours	0.73 mg/l	0.07 mg/l	Mortality Bioconcentration Behavioural and morphological changes	Substantial mortality and incidence of Behavioural and morphological alterations occurring at higher concentrations	significant positive correlation between tissue concentration and mortality with dose-dependent deformities and behavioural abnormalities. More pronounced poisoning symptoms at higher concentrations.	Ezemonye and Tongo, 2010a
		Bufo regularis (Toad)	0.01, 0.02, 0.03 and 0.04 µg/l	Adult	28 Days			Glutathione-S-trsnsferase (GST)	Alteration in Glutathione-S-trsnsferase (GST) activity in liver, brain serum, lungs, GIT tissues	GST is an efficient biomarker for Enosulfan toxicity	Ezemonye and Tongo, 2010b

D. LINDANE

	MOL-LUSCS	*Egeria radiata*	0, 50, 100, 150, 200, 250 (mg /l)	3.06-5.34cm (Length) and 48.0-89.5g (weight)	96 Hours	145 mg/l	14.5mg/l	Mortality Avoidance Response	Substantial mortality and incidence of abnormal avoidance response occurred at higher concentration	High mortality at day 4 indicated that the organism is sensitive to lindane. This can thus provides a rationale for monitoring of the waters of the Niger Delta ecozone	Okon and Akpan,1991

Histopathological Biomarkers used in Amphibian pesticides toxicology

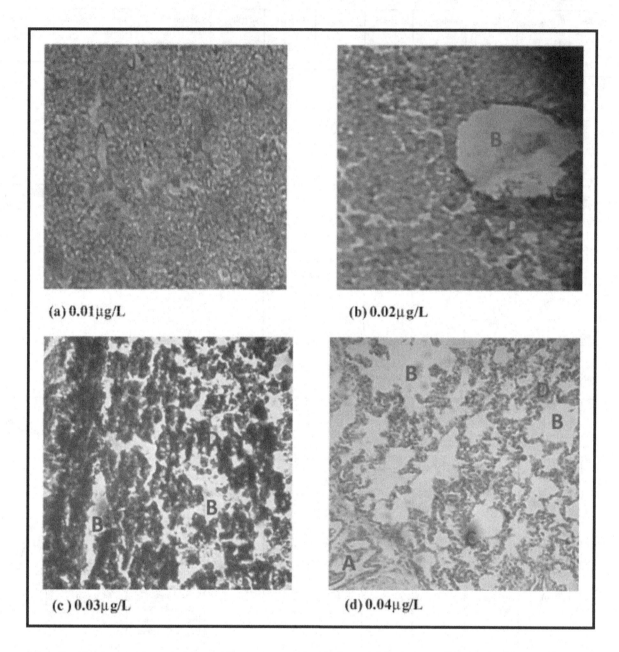

Plate 1: Photomicrograph of toad liver exposed to 0.01, 0.02, 0.03 and 0.04 µg/L Diazinon. (H & E stain x300). Dilation of sinusoids (A), cellular vacuolization (B) Pyknosis(C) cellular degeneration (D)and necrosis (E).

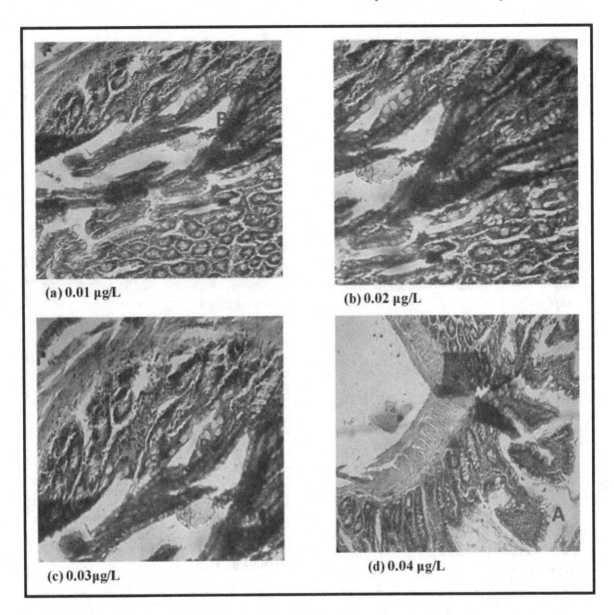

(a) 0.01 µg/L

(b) 0.02 µg/L

(c) 0.03µg/L

(d) 0.04 µg/L

Plate 2: Photomicrograph of toad GIT exposed to 0.01, 0.02, 0.03 and 0.04 µg/L Diazinon. (H & E stain x300). Severe hypertrophy in the muscularis (A), infiltration of mononuclear lymphocytes in the lamina propria (B), disintegration of epithelium layer (D)

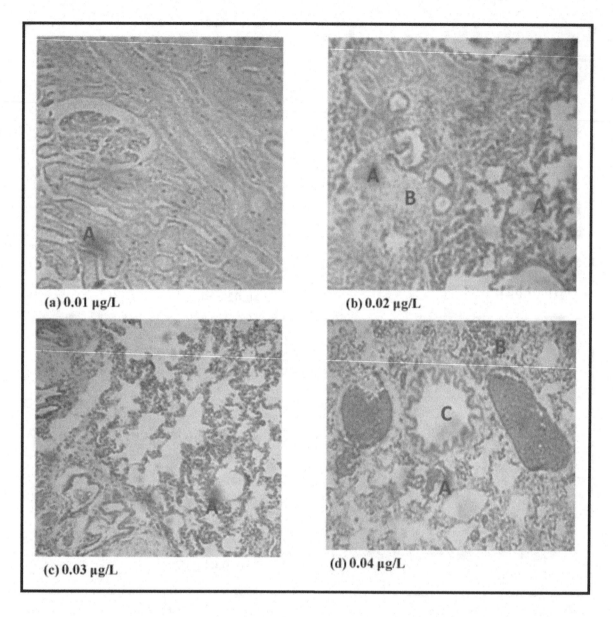

(a) 0.01 µg/L

(b) 0.02 µg/L

(c) 0.03 µg/L

(d) 0.04 µg/L

Plate 3: Photomicrograph of toad Lungs exposed exposed to 0.01, 0.02, 0.03 and 0.04 µg/L Diazinon (H & E stain x300). Area of interstitial fibrosis with grey intercellular staining of collagen (A), peribronchial lymphoid infiltration (B), alveolar fibrosis (C).

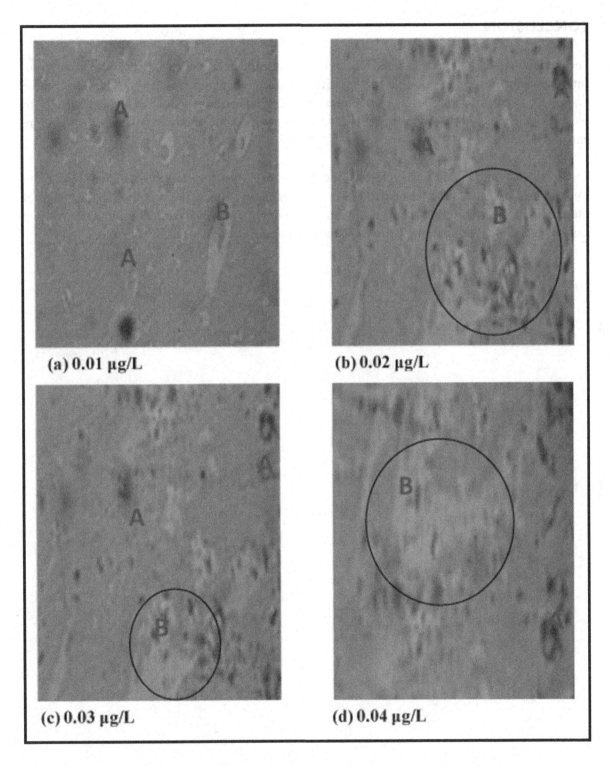

(a) 0.01 μg/L

(b) 0.02 μg/L

(c) 0.03 μg/L

(d) 0.04 μg/L

Plate 4: Photomicrograph of toad Brain exposed to 00.01, 0.02, 0.03 and 0.04 μg/L Endosulfan. (H & E stain x300). Dark-stained degenerating Purkinje neurons (A), Vacuolar changes with empty spaces which appeared as "moth eaten" areas (B).

CONCLUSION

The main goal of toxicological and ecotoxicological studies is to ensure that pollution from anthropogenic sources do not give rise to adverse effects on living organisms. Ultimately, these studies focus on measuring levels of pollution that may induce irreversible ecological changes to ecosystems. Biomonitoring of chemical pollutants on natural populations of organisms is therefore imperative.

Local researchers in Nigeria are therefore calling on the Government to establish a national information bank, to help collate researches carried out on Nigeria fauna. This will aid in the establishment of site-specific (Water, Soil, Sediment) quality criteria using the safe concentrations obtained, for the regulation and monitoring of these toxic chemical pollutants in ecosystems in Nigeria.

REFERENCES

1. Ezemonye L.I.N, Ogeleka D.F. and Okeimen F.E. (2007). *Desmoscaris tripsinosa* and *Palaemonetes africanus* Responses to Concentration of Neatex and Norust CR486 in Sediment. *Journal of Surfactants and Detergents*.No.10, pp. 301-308.

2. Ezemonye L.I.N, Ogeleka D.F. and Okeimen F.E. (2007). Acute Toxicity of Industrial Detergent (Neatex) and Corrosion inhibitors (Norust CR486) TO Early Stages of Ciclids: *Tilapia Guineensis*. *Journal of Chemistry and Ecology*. Vol. 23, No. 2, pp. 1-8.

3. Ezemonye L.I.N, Ogeleka D.F. and Okeimen F.E. (2007). Biological alterations in fish fingerlings (*Tilapia gunieensis*) exposed to industrial detergent and corrosion inhibitors. *Journal of chemistry and Ecology*. Vol. 23 (5), pp. 1-10.

4. Ezemonye L.I.N and Ilechie I. (2007). Acute and Chronic Effects of Organophosphate Pesticides (Basudin) to Amphibians tadpoles (Ptychadena bibroni). *African Journal of Biotechnology*. Vol. 6 (13), pp. 1554-1558.

5. Ezemonye L.I.N, Ogeleka D.F. and Okeimen F.E. (2007). *Desmoscaris tripsinosa* and *Palaemonetes africanus* Responses to Concentration of Neatex and Norust CR486 in Sediment. *Journal of Surfactants and Detergents*.No.10, pp. 301-308.

6. Ezemonye L.I.N, Ogeleka D.F. and Okeimen F.E. (2008). Lethal toxicity of industrial chemicals to early life stages of *Tilapia Guineensis*. *Journal of Hazardous Materials. No.157, pp. 64-68*.

7. Ezemonye L.I.N and Tongo I. (2009). Lethal and Sublethal Effects of Atrazine to Amphibian Larvae. *Jordan Journal of Biological Sciences*.2 (1):29-36.

8. Ezemonye L.I.N., Ogeleka D.F and Okieimen F.E. (2009). Lethal Toxicity Industrial Detergent on Bottom Dwelling Sentinels. *International Journal of Sediment Research*, **24,** 478-482. (Britain).

9. Ogeleka D.F., Ezemonye L.I.N and Okieimen F.E. (2010). Sublethal Effects of Industrial Chemicals on Fish Fingerlings (*Tilapia Guineensis*). *African Journal of Biotechnology*, **9** (12), 1839-1843. (South Africa).

10. Ezemonye L.I.N and Tongo I. (2010).Sublethal Effects of Endosulfan and Diazonin Glutathione-S-transferase (GST) in various tissues of adult amphibians (*Bufo regularis). Chemosphere*. 81:214-217

11. Ezemonye L.I.N and Tongo I. (2010). Acute Toxic Effects of Endosulfan and Diazinon Pesticides on Adult Amphibians (*Bufo regularis). Journal of Environmental Chemistry and Ecotoxicology*. Vol. 2(5): 73-78.

12. Ezemonye L.I.N., Olomokoro, J.O and Asuelimen, I.L. (2003). Evaluation of Toxicity of Some Industrial and Organic Effluent to Tadpole of *Bufo regularis* by Bioassay. *Journal of the Nigeria society for Experimental Biology (NISEB)*. **3,** (1), 13-22.

13. Ezemonye L.I.N., Olomokoro, J.O. and Negbenebor I.M. (2003). Evaluation of Toxicity of Laundry Effluent on Bio-indicators of Aquatic Pollution. *Nigerian Journal of Applied Science.* **21,** 56-61.

14. Dede, E. B. and Dogara, F. M. (2004). The acute Toxicological effects of Gammalin 20 on the lung and Pancreas of Guinea Pig. *J. Appl. Sci. Environ. Mgt.* **8** (1): 33-37.

15. Daka, E.R. and I.K.E. Ekweozor, 2004. Effect of size on the acute toxicity of crude oil to the Mangrove Oyster *Carasostrea gasar*. J. Applied Sci. Environ. Manage., 8: 19-22.

7.4   Sustainable Composting for Environmental Protection and Economic Development in Nigeria

Alamu, S. Abideen[i]

Sustainable Composting for Environmental Protection and Economic Development in Nigeria

Alamu, S. Abideen[1]

Absract

Nigeria is the largest country in Africa currently challenged with various waste management problems. On the daily basis, about 0.34 kg of waste per person is generated from domestic, agricultural and industrial activities in the country. The municipal authorities saddled with the responsibility of waste managment could not cope due to inadequate resources, poor technical capability and other logistics. While some inorganic wastes are scavenged and recycled into various products, large quantity of organic wastes have continued to be deposited on dumpsites, road sides and along major streets which latter become source of health and environmental problems to the people. Composting of organic wastes into soil conditioner is a strategic way of reducing the menace of organic wastes in the country. This paper seeks to examine the sustainable composting of organic waste in Nigeria. Desk research on sustainable compost production and associated benefits of compost provides data for the paper. Findings suggest that successful production and marketing of compost will help to reduce the volume of organic in the environment, produce organic fertilizer, reduce pollution, create jobs and promote environmental sustainaility in the country. The study suggests short and long term measures that will make sustainable composting in Nigeria a reality.

Key words: Environment, sustainable composting, organic waste, Nigeria

---

[i]   The author is a Research Fellow in the Technology Development Division of the Nigerian Institute of Social and Economic Research (NISER), Ibadan, Nigeria.

List of Abbreviations

CEC   Cation Exchange Capacity

EU   European Union

MSW   Municipal Solid Waste

MT   Metric Tone

NISER   Nigerian Institute of Social and Economic Research

N and P   Nitrogen and Phosphorous

NBS   National Bureau of Statistics

UI   University of Ibadan

UK   United Kingdom

PH   Degree of Alkalinity and Acidity

PAH   Polycyclic Aromatic Hydrocarbon

PCB   Polychlorinated Biphenyls

S&T   Science and Technology

1   Introduction

Nigeria is the largest country in Africa having a population of about 140 million people with an annual population growth rate of 2.38% (NBS, 2008). It is a developing country challenged with various environmental problems. The major environmental problems in Nigeria include erosion (gully and sheet), flooding, biodeversity loss, climate change, deforestation and poor waste management (Adeyinka et al 2005). The large size of Nigerian population coupled with various agricultural, industrial and domestic activities as well as consumption pattern of the people are responsible for the generation of large quantity of waste in the country. In addition, consumption of large quantity of non-durable and poor quality products also contribute to the increasing rate of waste generation across the country. Sridhar (2007) estimates that about 0.34kg waste per person is generated in Nigeria on daily basis. The amount of waste generated per head in the country is expected to increase due to population increase, goods consumption pattern of the people and identification of wastes as burden.

Waste will continue to be a major environmental problem in most of the urban cities in Nigeria due to lack of effective waste management plan. At present, there is no effective plan for the recycling of the available wastes into useful products. In addition, the available waste materials are regarded as nuisance to the environment and are dumped haphazardly on any available place. Wastes dumped in unwanted places have continued to pollute the environment and also become source of diseases and breeding place for rodents and diseases vectors. On many occassions, smoke produced from burning of wastes usually pollute the environment and make life uncomfortable for the people.

Municipal solid waste (MSW) which include refuse from households, non-hazardous solid waste from industrial, commercial, institutions, market and so on is the largest type of waste generated in the country. The MSW stream generally consists of inorganic/ non-biodegradable and organic / biodegradable wastes. The inorganic wastes include materials like metals, glass, plastics and a host of others while the organic components include domestic food waste, crop residues, wood waste, garden waste, livestock dung and poultry droppings that are rich in organic matter, nitrogen, phosphorus, potassium and micronutrients.

The size of organic waste often constitute more than half by weight of the total solid waste generated in Nigeria and also in many countries across the globe. For example, organic waste stream in most cities in Nigeria has been estimated to be higher than the inorganic counterpart. It has been estimated that MSW generated in major cities in southwest Nigeria consists of garbage (60.5%), paper (19.1%), sand (9.3%), plastic (7.1%), glass (1.7%), and metal scrap (1.8%) (Adepetu et al 1999). In some developing countries, the percentage of organic matter often outweigh the inorganic matter. For example, Katzir (1996) stated that the percentage of organic materials in city wastes in developing countries reaches up to 80 percent of the total waste generated.

Management of organic waste in urban places in Nigeria is a major challenge confronting the municipal authorities due to their inadequate resources, poor technical knowhow and poor political will. In many occassions, local authorities in Nigeria spend more than 70% of their revenue on waste management (Ogwueleka, 2003), but could only collect about 50 % of the available waste materials. In addition, wastes are not converted into useful products due to poor technological know-how of the municipal authorities. The failure of the resource-poor local authorities to collect and recycle the organic waste into useful products always leads to unpleasant condition of the major cities in Nigeria. As a result, the uncollected organic wastes decompose and pollute soil and water resources and constitute serious health problem to the people. As a result, waste problem has emerged as one of the greatest environmental challenges facing the country.

In most of the urban cities in Nigeria, an appreciable quantity of the inorganic wastes are collected and recycled for the production of various types of plastic and metal products. For example, wastes scavengers are seen along major streets and dumpsites in the country collecting plastic, metals and other inorganic wastes for recycling into various domestic and industriial products. It is however worrisome, to note that organic wastes that constitute the bulk of total available wastes are abandoned on dumpsites, on major streets and along the road side to become source of environmental and human health problems. Organic wastes are highly putrescible, favour breeding of flies and harbour infectious diseases like Tetanus, Thyphoid, Cholera, Dysentary and others. The leachates from these wastes promote mosquito breeding, contaminate water bodies and irrigable land. The solid component of organic wastes also destroys the aesthetic value of the environment, harbours rodents and produces foul smell and therefore causes unpleasant city condition, serious health and environmental hazards. The solid components of organic wastes also release methane, a green house gas (GHG) into the atmosphere during decomposition.

The fact that organic waste produces methane, contaminates ground water, harbours diseases vectors and produces offensive odour shows that it contributes negatively towards making our environment sustainable. More importantly, management of organic wastes is a major problem confronting most of the urban cities in the continent of Africa. Therefore, proper management of organic waste is an important component of environmental sustainability measures across the globe.

In order to curtail the various environmental consequences of organic waste and achieve sustainable environmental resources management in Nigeria, there is the need to recycle the organic wastes into soil conditioner through composting. This will reduce the release of bad odor, occurrence of flies, as well as contamination of the underground aquifer by nitrate, heavy metals and bacteria caused by the presence of organic wastea in the environment. The successful composting of large quantity of organic wastes in the country into useful products will obviously make a great impact on the economic and social development of the people and protect the environment.

The objective of this paper is to examine the possibility of sustainable recycling of organic wastes into soil conditioner via composting for achieving environmental protection and economic development in Nigeria. The paper looks at successful composting as a strategic measure for achieving reduction of the various environmental problems associated with the presence of organic waste in the Nigerian environment as well as the spill-over effects on the economic development of the country.

2   Materials and Method

This is a policy oriented research paper produced through desk research from Library and the Internet search. Desk research on wastes problems, recycling of organic wastes and sustainable composting in Nigeria and some other countries across the globe provides data for the paper. The information used in the preparation of this paper is obtained from various journal articles, government publications and international periodicals. Content analysis of the information obtained was principally employed in the preparation of the paper.

Results and Discussions

Environmental Sustainability and Composting

Prior to 1992, the various adverse effects of human socio-economic activities manifested as major threats to the air, water, soil, plant and other resources of the Earth. The alarming rate of destruction and pollution of natural resources was observed as clog in the utilisation of these resources by the future generation if the trend continues. This led to the convergence of the world leaders in 1992, to revert the ugly situation in order to ensure sustainable utilisation of natural resources.

At the Earth Summit in Rio de Janerio in 1992, the world leaders signed a global environment and development action plan called *Agenda 21.1*(EU, 1997). Since then, sustainable development has emerged as an important issue to the global community in their development planning. Parts of the strategies for achieving a sustainable environment is the recycling of organic waste. One major way for achieving reduction in the amount of biodegradable municipal waste going into landfill and other disposal sites is composting of organic wastes as stipulated in Article 5(2) of European Union (EU) directive. In order to comply with this directive, the United Kingdom (UK), for example, published a policy framework for ensuring increase in the quantities of organic material that is recycled through composting (UK, 1999). Other countries also comply with the resolution of the Earth Summit. For example, Israel has put strong regulation in place that ensured an increased in the amount of organic waste recycled from 48,000 MT in 1998 to 200,000 MT in 2004 (Kan et al, 2008). Composting is basically a process for decomposition of organic solid wastes by various microorganisms like bacteria, actinomycetes and fungi to produce relatively stable humus organic end product, carbon dioxide, water and energy. It is one of the most efficient and environmental friendly methods of solid waste disposal with several advantages when compared with landfill disposal on which many countries across the globe currently depend. It is an age-long practise in many rural areas, where farmers traditionally put plant and animal wastes on their fields to enhance soil fertility and reduce the available waste stream.

The Benefits of Composting Organic Wastes

Composting remains the most viable and environmental friendly method of managing organic waste throughout the world (Iyeker, 2006). The method has several economic and environmental benefits over other waste managemnet strategies such as landfill, waste incineration and disposal to water bodies. Some of the benefits of composting are itemised below:

Offers a Better Waste Management Strategy; Recycling of organic waste to compost provides a useful management measures over incineration and disposal to water bodies. Composting of organic wastes obviously reduces several health and environmental problems caused by indiscriminate dumping of organic wastes into streams and road side. This invariably prevents obstruction of highways and blockage of streams by waste materials.

Reduces Quantity of Waste for Disposal: Composting turns putreceable waste into resources. As a result of composting, organic waste becomes raw material for compost manufacturing plants. The available organic waste would therefore be sorted and transported to compost manufacturing factory. Composting therefore, helps to reduce the quantity of waste available for disposal and incineration.

Converts Organic Matter into Safer and Useful Product: Compost is a useful and safer products for soil fertility improvement and a remediator for soil polluted with organic and inorganic pollutants. At times, compost are blended with mineral fertilizer for soil fertility improvement. Presently, the adverse effects of compost in the environment is unknown

Recycles Nutrients and Enhance Long Term Soil Fertility Maintenance: Compost improves soil structure and porosity for better plant growth, ease root growth, reduces run-off and erosion. Compost also buffers soil degree of acidity and alkalinity (PH) and improves the soil cation exchange capacity (CEC). A study conducted by Golabi et al (2006) on effects of compost on soil structure and fertility improvement shows that land application of organic compost enhanced soil quality, increased soil fertility level and improves the soil bulk density, organic matter content, nutrient distribution and other soil quality parameters. Beltram et al (2002) stated that composted organic material contains Nitrogen and Phosphorous (N and P) which are steadily available for plant growth due to slow release of these essential plant nutrients. Organic matter serves as a very important source of plant nutrients and micronutrients may be satisfactorily supplied by decomposing organic matter.

Provides Income and Employment: Composting provides employment and income for the poor. The various processes of handling waste such as separation, segregation, transportation, shredding, mixing and so on provide job opportunities and income for the poor.

Controls Environmental Pollution: Composting has several environmental benefits. For example, Nyamangara et al (2003) stated that the use of composted organic waste as fertilizer and soil amendment does not only results to economic benefit to the small-scale farmers but also reduces pollution due to reduced nutrient run-off and nitrogen leaching. Various studies have shown that compost has the ability to remediate soils that are contaminated by heavy metals (Mule and Melis, 2003). Other studies have equally confirmed the remediating capacity of compost on various matter. For instance, compost has been used for remediating contaminated soil (Williams and Keehan, 1993). Study conducted by Barker and Bryson (2001) also confirmed that compost remediates both organic and inorganic pollutants such as pesticides, polycyclic aromatic hydrocarbons (PAHs), and polychlorinated biphenyls (PCBs).

Reduces Problems Associated with Mineral Fertilizer Consumption: Science and Technology (S&T) has enhanced a wider understanding of plant and soil chemistry and also led to improved fertilization and farming practices that have increased crop yields worldwide (Tisdale et al 1985). However, over-application of mineral fertilizers has been reported to reduce farm profits, cause soil degradation and environmental pollution (Tisdale et al, 1985. The ease of applying synthetic fertilizers and the lack of knowledge for matching fertilizer applications with the nutrient requirements of certain crops have added to the problem (Golabi et al, 2006). However, Hue (1992) reported that green manures and composted organic material increase soil organic matter and provide plant nutrients, alleviate aluminum toxicity, and render phosphorus more available to crops. In places where adequate demand and supply of compost exist, the rate of consumption of mineral fertilizers and attendants environmental consequences are considerably reduced.

Feasibility of Compost Production in Nigeria

The necessary input, technology and human resource for sustainable composting are available in the country. Some states government in the country have already established compost manufacturing plants in their states' capitals. For instance, compost manufacturing plants are already established in some major cities in the country like Ibadan, Akure and Lagos. In Akure, for example, machines have been fabricated for the recycling of organic waste to compost and for mixing the compost with certain percentage of mineral fertilizer, with daily production of 5 tons (Olanrewaju and Ilemobade, 2009). Similar compost production plant is also in Bodija market, Ibadan, Nigeria.

Technology Requirement for the Production of Compost

Various types of technologies are used for the production of compost. They include barrels and simple shreadders, large windrow-turning machine. The chioce of technology depends on composter's capacity to operate and maintain, labour cost, power input cost, production capacity and nature of the waste. However, Nigeria has the technological capability for fabricating and maintaining compost manufacturing plant.

3   Market Requirement for Compost Products

Sustainability of compost production depends largely on continous sale. On the demand side, are the farmers who need it for maintaining their soil fertility. A very large proportion of Nigerians are crop producers who mostly depend on mineral fertilizers for crop yeild improvement. The fact that Nigeria is an agrarian economy shows that the economy can absorb huge tonnage of compost produced in the country. In addition, Nigeria is one of the largest importers of fertilizer for crop production purposes in the sub-Sahara Africa. Sky would be the limit in the marketing of compost in the country if there is breakthrough in the awareness of the farmers on utilisation of compost for soil fertility maintenance,

Input Requirement for the Production of Compost: The major raw material for compost production is organic waste. On the daily basis, large amount of organic wastes are generated from various industrial and domestic activities across the country. Therefore, ample supply of organic waste for feeding compost manufacturing plants can readily be guaranteed in the Nigerian economy.

Challenges in the Production of Compost in Nigeria

Small Size of Compost Market: Sustainability of a commercial compost plant in Nigeria will depend on continous sale of the compost. For instance, if large quantity of household waste are converted to waste, automatically, large market will be needed. As at now, the current known market for compost seem fairly restricted to household gardeners, nurseries and organic farmers. As a result, the percentage of compost users in the country are limited and so there is need for the expansion of customers' base through awareness building and identification of new users

Risk of Investing on Compost Manufacturing: The small size of compost consumers is an impediment to investing in composting. Most of Nigerian investors may consider composting to be an unviable venture because of the risk involved, product marketability and their eagerness to make profit quickly. As a result, composting may not be attractive to Nigerian investors.

Problem of Re-orientation of Farmers on the Use of Compost: Nigerian farmers mostly depend on mineral fertilizers for their soil fertility improvement. The situation in the country is that largest share of the farmers grossly depend on mineral fertilizer for their soil fertility maintenance. As a result, it may take some time before these farmers will adopt compost because of their lack of knowledge about the efficiency of compost as an effectiev and efficient soil conditioner.

Stigmatization of Working in Compost Factory: A potential job seeker in Nigeria prefers working in decent environment such as factory or government establishment rather than working in a dirty environment like compost factory. People working in compost manufacturing firm will be dealing with handling of dirty organic waste materials that produce odour. As a result, working in a compost factory is often considered to be unpleasant, embarrasing and degrading. It may therefore be difficult to recruit qualified workers in compost factory in Nigeria unless the working environment is made attractive.

4   Recommendations and Conclusion

With regard to the above mentioned challenges, it may be difficult for achieving sustainable composting in Nigeria. In order to realise sustainable composting in the country, there are short and long terms measures to be put in place in waste management policy.

Short Term Measure: At present, it would be difficult for an individual, or a non-governmental organization to embark on large scale compost production because ver few people are using it for soil fertility improvement. As a result, any compost factory set up by an individual or a private establishment will soon collapse due to small size of market for compost. It is therefore advisable for government, most espacially state or local to take the lead due to the following reasons:

Production of High Quality Compost: If government kickstart the production of compost, it is assumed that production of good quality compost will be ensured. Compost produced by the government must conform with local or international standards as government would have its own image to protect through ensuring product standardization.

Effective Marketing Strategy: Again, an effective marketing strategy will be needed to increase consumers' demand for compost. There is need for popularising compost for soil fertility maintenance in the country. This can be achieved through placement of regular advertisement on the utilisation of compost by farmers and home gardeners for soil fertility improvement on the prints and electronic media as well as interraction with the farmers through the agricultural extension agents.

Provision of Funds: Adequate funding of compost plant will be guaranteed under the control of government most especially at the initial stage. The price of compost must be made affordable to the famers through subsidy or other means. Profit-making should not be the topmost priority agenda of the venture most especially at the initial stage.

Better Condition of Service: Good remuneration in terms of payment of attractive salary and allowances and provision of other better condition of service for the workers will surely motivate them to work in compost factory. This will reduce stigmatisation of working in compost factory by qualified workers

Long Term Measures: Individual and private organisation must be encouraged to set up compost manufacturing plants in the country as soon as compost is successfully popularised among the Nigerian farmers,. This can be achieved through granting loans and other incentives to individuals and other organisations in the country that are interested in going into composting. In addition, source collection of organic waste materials should be encouraged in the country. Government should put appropriate legislation in place that will compel compost manufactoring firms to pay for already sorted organic waste materials.

More importantly, the habit of indiscriminate disposal of organic wastes into any available place should be abolished with strict penalty in the country. Strict penalty should be awarded against waste disposal on our highways, illegal dumpsites and major streets in the country. Every Nigerian must be sensitised about the importance of environmetal sustainability and the need to achieve it. The Nigerian waste manegement plan should be geared towards reduction in the amount of organic waste generated across the country. The amount of organic materials such as food and crop waste should be reduced in the country. Government should intensify efforts to reduce the amount of crop materials that are wasted every year, especially during harvesting periods through devising better processing and storage techniques for perishable crops in the country.

REFERENCES

Adeyinka, Adewunmi; Bankole, Phillips. and Solomon, Olaye. (2005). Environmental Statistics: Situation in the Federak Republic of Nigeria. Being country report presented at the workshop on environmental statistics held in Dakar, Senegal, 28th February and March 4th

Barker, Allen and Bryson, Gretchen (2001). Bioremediation of Heavy Metals and Organic Toxicants by composting. *The Scientific World Journal 2, 407-420.*

Beltran E M, Miralles de Imperial R, Porcel M A, (2002). Effect of Sewage Sludge Compost Application on Ammonium Nitrogen and itrate-Nitrogen Content of an Olive Grove Soils. A Proceedings of the 12th International Soil Conservation Organization Conference Beijing: Tsinghua University Press, 2002. 395-403.

European Commission (1997). Agenda 21 The First Five Years, p 5.

Golabi, Mohammed.; Denney, Peggy and Iyekar, Clancy. (2006). Composting of Disposal Organic Wastes: Resource Recovery for Agricultural Sustainability. *The Chinese Journal of Process Engineering vol.6, No.4.*

Hue N V (1992). Increasing Soil Productivity for the Humid Tropics through rganic Matter Management [A]. Tropical and Subtropical Agricultural Research, Progress and Achievements [C]. The Pacific Basin Administrative Group, University of Hawaii and University of Guam.

Kan, Iddo; Ayalon Ofira, O and Federman, Roy (2008). Economic Efficiency of Compost Production: The Case of Israel. Discussion Paper N0,8.08

Katzir R. 1996. Workshop on Market Gardening, Farm Associations, and Food Provision in Urban and Peri-Urban Africa. Netanya, Israel, June 23-28,1996.

P Mule` and Pietro Melis (2003). Community Soil Science Plant Analysis, Vol. 31 pp 3193.

National Bureau of Statistics (NBS) (2008). Annual Abstract of Statistics

Nyamangara, J, Bergstrom L F, Piha M I, (2003). Fertilizer Use Efficiency and Nitrate Leaching in a Tropical sandy Soil. *Journal of Enivironmental Quality Vol 32: 599-606.*

Ogwueleka, Chuks., (2003). Analysis of urban solid waste in Nsukka, Nigeria. *Journal of Solid Waste technology and Management, 29 (4): 239-246.*

United Kingdom (UK) (1999). Department of the Environment, Transport and the Regions. A Way with waste Olanrewaju, Olawale and Ilemobade, Albert (2009). Wastes to wealth:A case study of Ondo state integrated waste recycling and treatment project, Nigeria. *European Journal of Social Sciences, Volume 8 Number.1*

Tisdale Samuel, Nelson Werner, Beaton James (1985) Soil Fertility and Fertilizers, Fourth Edition [M]. New York: Macmillan Publishing Company, pp 112.

Sridhar, Mynepalli. (2007). Climate Change and Waste Management.

Williams, Richard and Keehan, K.R. (1993) Hazardous and industrial waste composting. In *Science and Engineering of Composting*. Hoitink, H.A.J. and Keener, H.M., Eds. Renaissance Press, Worthington, OH. pp. 363-382.

7.5 The Effect of Poultry Manure And Chemical Fertilizer on the Growth of chrysophyllum albidum Seedlings

[1]*Ojo, O.S; [2]Ojo, O.A; [1]Akinyemi, O; [1]Sodimu, A.I; Suleiman, R.A[2]

Abstract

For twelve weeks, poultry manure and NPK 20:10:10 fertilizers were experimented on the growth of *Chrysophyllum albidum* seedlings. A Complete Randomized Block Design (CRBD) was used with a layout of seven treatments and each treatment was replicated four times. The result obtained in this study showed that *Chrysophyllum albidum* seedlings positively responded to both fertilizers, but it is best in organic manure. Seedlings supplied with 15g of poultry manure significantly ($P<0.05$) produced best morphological parameters in seedlings of Chrysophyllum *albidum*. The best morphological parameters obtained were, plant height 18.7cm, leaf number 8 and the stem diameter 0.36cm. The fresh weight and oven dry weight were also significantly superior to other treatments i.e. fresh weight was recorded as 4.12g and dry weight was recorded as 2.80g. There was also a significant difference among the treatments in terms of height, while leaf count and stem diameter were not significant at 5% probability level when subjected to Analysis of Variance (ANOVA). It is recommended that Government should subsidize the cost of inorganic fertilizer, and that nursery workers should use poultry manure as a component with nursery soil because it is a cheap source of fertilizer to enhance the production of quality seedlings of *Chrysophyllum albidum*.

Key words: Forestry, Sustainability, Manure and Development

[1]* Ojo, O.S; [2]Ojo, O.A; [1]Akinyemi, O; [1]Sodimu, A.I; Suleiman, R.A[2]

[1]   Department of Forestry Technology, Federal College of Forestry Mechanization

[2]   Forestry Research Institute of Nigeria P.M.B 2273, Afaka Kaduna, Nigeria

[2]   FRIN/JICA, P.M.B. 2010, Afaka, Kaduna,

*   **Corresponding Author** *lesluv247@yahoo.com*

1   Introduction

There has been much controversy over organic versus inorganic fertilizers. It is important to realize that plants do not recognize the difference between organic and inorganic fertilizers. Their tiny root hairs can absorb only nutrients that have been broken down into inorganic, water soluble forms. The judicious management and conservation of the soil to guide against these problems that eventually lead to decreased crop yield under intensive cropping have become major areas of agronomic research (Brechin and Mc Donald, 1994). The use of inorganic fertilizer has not been helpful under intensive agriculture because it is often associated with reduced crop yield, soil acidity and nutrient imbalance (Kang and Juo, 1980; Obi and Eso, 1995; Ojeniyi, 2000). Soil degradation which is brought about by loss of organic matter accompanying continous cropping becomes aggravated when inorganic fertilizers are applied repeatedly. This is because crop response to applied fertilizers depends on soil organic matter (Agboola and Omneti, 1982). The quality of soil organic matter in the soil has found to depend on the quality of organic materials which can be introduced into the soil either by natural returns through roots, stubble, slough off roots nodules and roof exudates by artificial application in form of organic manures which can otherwise be called organic fertilizers. The need to use renewable form of energy and reduced cost of fertilizing crops has revived the use of organic fertilizers worldwide. Improvement of environmental conditions and public health are important reasons for advocating increased use of organic materials (Seifritz, 1982; 2000; Maritus and Vlelc, 2001). The benefits derivable from the use of organic materials have, however, not been fully utilized in the humid tropics partly due to the huge quantities required in order to satisfy the nutritional needs of crops, transportation as well as the handling costs which constitute major constraints. Complementary use of organic manures and mineral fertilizers has been proved to be a sound soil fertility management strategy in many countries of the world (Lombin et al.,1991). High and sustained crop yield can be obtained with judicious and balanced NPK fertilization combined with organic matter amendment (Kang and Balasubramanian, 1990). A system integrating different practices of soil fertility maintenance is required and this will include the use of mineral fertilizer, organic manures and intercropping which provides a fast and good ground cover and also allows the roots to exploit soil nutrients at various depths (Steiner, 1991). There are however, advantages and disadvantages to each form of fertilizer, organic and inorganic.

1.1   Description

It is a small to medium buttressed tree species, up to 25-37 m in height with a mature girth varying from 1.5 to 2 m. Bole is usually fluted, frequently free of branches for 21 m. Leaves are simple, dark green above, oblong-elliptic to elongate obovate elliptic, 12-30 cm long and 3.8-10 cm broad. Fruits almost spherical, slightly pointed at the tip, about 3.2 cm in diameter, greenish-grey when immature, turning orange-red, yellow-brown or yellow, sometimes with speckles.

1.2   Uses oF *C. albidum*

The fleshy and juicy fruits, which are popularly eaten, are the potential source of a soft drink. Wood **is** brownish-white, soft, coarse and open in grain; very perishable in contact with the ground**.** Easy to saw and plane, nails well. It takes a fine polish, and therefore is suitable for construction work, tool handles, furniture and similar purposes. The fruits can be fermented and distilled for the production of wine and spirits and the common name is African White star apple. It is a tree of the rainforest. It has been found very useful in West Africa as a timber tree and the fruit as a source of food. Akinyele and Keshinro (1974) reported that *Chrysophyllum albidum* fruits contain high amount of vitamin C. *Chrysophyllum albidum* fruit has 8% protein 17.% oil, 21% sugar 584 Afr. J. Agric. Res. and 11% starch. The fruit pulp has been used on experimental basis for excellent jams, jellies with good appearance and flavor. *Chrysophyllum albidum* fruits are used to make wine in South Africa and in Zambia, alcoholic beer from the astringent fruits have been brewed.

1.3 Organic fertilizer

1.4 Advantages

Organic nutrients include things as cow, sheep, and horse manure (one should avoid using pig, dog or cat feaces because of the problems involved with internal parasitic worms which may be transferred to human). Bone meal, blood meal, compost and green manures will also provide nutrients for plants. There is less danger of over-fertilization by adding decomposed organic material to a plant. It provides a slow release of nutrients as microorganisms in the soil break the organic material down into an inorganic, water-soluble form which the plant can use. The addition of organic material; improves soil structure or "work-ability" immensely. It has been observed that addition of manure increases soil water holding capacity and this means that nutrients would be made available to plants where manure has been added to the soil (Costa et al, 1991).

1.5 Disadvantages

For the most part, organic fertilizer is not immediately available to the plants. As noted above, the "slow-release" feature can be an advantage. However, if there is an immediate need for nutrients, organic fertilizer can not supply them in a hurry. Furthermore, information on the amount of nutrient and the exact element in an organic fertilizer such as manure is not readily available to the farmers.

1.6 Inorganic fertilizer

1.7 Advantages

The primary advantage of using inorganic fertilizer is that nutrients are immediately available to the plants. As well, the exact amount of given element can be calculated and given to plants.

1.8 Disadvantages

Inorganic fertilizer, especially nitrogen, is easily washed below the level of the plant root through the leaching of rain or irrigation. An application which is too heavy or too close to the roots of the plants may cause "burning" (actually, a process of dessication by the chemical salts in fertilizer). If organic materials are readily available and cheap, the expense of the organic fertilizer should also be considered.

1.9 Medicinal importance

The medicinal uses of *Chrysophyllum albidum* appears to be limited to its natural habitat in the lowland rainforest belt of Nigeria; otherwise, the fruit is virtually the only part of the plant that is very commonly known and eaten across the whole country from the southern to the northern parts. The powder obtained from the bark is used locally in the treatment of bronchitis. The trees, which are usually tall, about 25m to 35m high are perennial and can be found in farms and homestead gardens. The bark, which is usually pale grayish brown in colour, is used in the preparation of medicine for the treatment of fever and black coated tongue called "Efu-dudu" in Yoruba. The powder obtained from the dried root is used locally in the treatment of rheumatism. The leaves are very important for various medicines. It is usually cooked with other herbs to make anti-malaria drugs. The leaves in combination with pure honey are used in the treatment of dry coughs. When the same preparation is used as body massage, it aids the nerves for better sexual performance and becomes very useful in the treatment of frigidity in women and depressed libido in men.

2   Objective of the study

The general objective is to assess the effect of poultry manure and chemical fertilizer on the growth of *Chrysophyllum albidum* seedlings.

The specific objectives are:

1. To determine the influence of different fertilizers on the growth of *Chrysophyllum albidum* seedlings using poultry manure and NPK.
2. To examine the effect of various levels of application of poultry manure and NPK on the growth of *Chrysophyllum albidum* seedlings.

2.1   Justification

Trees and shrubs could be seen as the source of many products besides timber. These products make an important contribution to the welfare and quality of lines of urban as well as rural communities. Much of the knowledge and technology for processing medicinal plants is only available in rural communities and only prepared by words of mouth and within families and small communities (Flores Rodas, 1986). The non cultivation of *Chrysophyllum albidum* by both the Government and private sectors makes it necessary to encourage people to cultivate it because of its many uses. For the species present and future demand requires appropriate medium required for the growth of seedlings in the nursery. The population of Nigeria is increasing just like the world's population and there is high demand of wood products in the present day, hence emphasis needs to be on the area of establishing indigenous plantation of *Chrysophyllum albidum*.

3   Materials and Methods

The experimental design was a layout in a Complete Randomized Block Design (CRBD) and table was included at the glossary consisting of each treatment contain three pots, the treatments were replicated four times as below;

Factor 1 -   Poultry Manure (P/M)

Factor 2 -   Chemical Manure (NPK)

Factor 3 -   Control (c)

The seedlings were three weeks old at the point of collection with equal height and number of leaves to one another. Transplanting of seedlings was done into polythene pots containing treated soil containing sundried poultry manure at the rate of one seedling per pot.

NPK 20:10:10: fertilizer was applied a week after transplanting the seedlings. The initial measurement were taken after transplanting and watered daily. Records of growth were observed fortnightly until the termination of the experiment after twelve weeks of observation.

Parameters observed include plant height measured from the base of the plant above the soil level to the terminal bud of the plant, stem diameter was measured at 2cm above the soil level with calipers. The numbers of leaves were recorded. At the end of the experiment, plant parts were oven dried at 76°C for 36 hours to obtain the total dry yield.

4   Results

The table 1 shows that 15g poultry manure produced seedlings that were significantly superior to the seedlings in other media whose height is tallest with 13.7 cm.

The data on height parameters as shown in above were subjected to Analysis of Variance at less than 0.05 level of significance and the result indicated a significant difference.

Table 2 also show the values obtained in the stem measurement in the course of the experiment. The study shows that there is a significant change in the stem diameters with an increase in time.

Similarly the experiments were subjected to analysis of variance to further show the significant relationship between the poultry manure and the parameters used. At 95% percentage probability the experiment revealed significant having proceeded to LSD with 0.29039 for the height and 0.00116 for the stem measurements.

4.1   Discussion

The data indicated in this research implies poultry manure has considerable influence on the growth of *chrysophyllum albidum* seedlings especially where the treatment containing poultry manure and NPK fertilizer would be appropriate potting mixture for *Chrysophyllum albidum*. The treatment with poultry manure showed a gradual increase in height from the 2nd week after transplanting than the treatments with NPK fertilizer and the control experiment respectively. After the fourth week, treated plants of NPK fertilizer showed rapid growth on *Chrysophyllum albidum* seedlings height. This might be due to rapid mineralization of inorganic fertilizer over manure. Ojeniyi (2000) stated that the use of inorganic fertilizer has not been helpful under reduced crop yield, because it is often associated with reduced crop yield, soil acidity and nutrient imbalance.

The level of application of NPK enforces it to remain penultimate because of its toxic nature to plants as this study shows that the increment in the level of poultry manure application should be encouraged. Therefore, this study observed that *Chrysophyllum albidum* responded best to 15g of both poultry manure and NPK 20:10:10: fertilizer. Growth responses were generally better in seedlings grown with organic manure than those grown with inorganic fertilizer. Among the seven treatments those grown with 15g of poultry manure recorded higher values in morphological parameters.

5   Conclusion

Generally, seedlings growth responded to organic manure than inorganic fertilizer. The quantity of soil organic matter in the soil has been found to depend on the quantity of organic material which can be introduced into the soil by application of organic manures which can otherwise be called organic fertilizers (Maritus and Vlelc, (2001). This might be due to increased water holding capacity created by organic manure media. This confirmed as the recognition of organic and inorganic fertilizer as effective silvicultural tools for raising plants in the nurseries and for harvesting the growth of plants (Sara, 2005).

5.1   Recommendation

Poultry manure as a means of adding nutrients into the soil for plant use is a major silvicultural tool and helps the plant to grow optimally. It is recommended that seedlings of *Chrysophyllum albidum* should be raised with the addition of poultry manure at high level so as to achieve a better result. Poultry manure is useful in the fertility of the soil, prevent soil erosion and prevent leaching of soil nutrient, also, cost of inorganic fertilizer should be subsidized by government so that most nursery workers could be afforded. Farmers should also embark on the use of this cheap, available, effective and suitable poultry manure in their business to reduce the demand of inorganic fertilizer.

Apparently, there is need for further investigation on types of inorganic materials that are readily available in order to prepare a check list of the appropriate potting mixtures of different inorganic materials. Also, the combined effect of both organic and inorganic fertilizers should be looked into so as to know the combination of both fertilizers can be the best option for achieving optimal growth of *Chrysophyllum albidum* seedlings at the nursery state in Nigeria.

Nevertheless, poultry waste can be managed to reduce the uncontrolled waste deposit in surroundings of farmyard. It is also important to note that poultry manure will be used to grow all categories of crops without any threat of damage to plant product quality. Finally, it helps to minimize the effect of continuous cropping on piece of land by almost neglecting shifting cultivation practice.

REFERENCES

Agboola, A.A and Omueti, J.A (1982). Soil fertility and its management in tropical Africa. Paper presented at the Institute of Tropical Agriculture, Ibadan, Nigeria. P. 25.

Akinyele, I.O and Kenshinro, O.O. (1974). Tropical fruits as source of vitamin C. food Chem. 5: 163-167.

Anazonwu, J.N. (1981). Indigenous food and nutritional a. Ministry of Science and Technology, Enugu, Nigeria p. 50.

Brechin, J. and McDonald, G.K. (1994). Effect of form and rate of pig manure on the growth and nutrient uptake and yield of Barley (Co. Gallen). Australian J. Exp. Agric., 34: 505-510.

Cook, G.W. (1979). Same priorities for British soil science 30: 187-213.

Flore Rodas M.A. (91986). Some medicinal plants in Africa and Latin America, F.A.O. of United Nations, Rome. Pp. 43-47.

Kang, B.T. and Balasubramanian, V. (1990). Long term fertilizer trial on afisols in West Africa, in transaction of XIV International Soil Science Society (ISSS) congress, Kyoto, Japan. Pp. 350.

Keau, R.W.J. (1989). Trees of Nigeria. A revised version of Nigeria Trees (1960, 1964). Oxford Science Publication, pp. 102.

Labode, P (2003). Forests, people and the environment. Proceedings of a National Workshop organized by FANCONSULT and Edo State Chapter of forestry Association of Nigeria held in Benin City, Edo State, 5-6 September, 2002. Pp. 26-40.

Lombin, L.G.: Adepetu, J.A. and Ayotade, A.k. (1991). Organic fertilizer in the Nigerian Agriculture: Present and future F.P.D.D. Abuja. Pp. 146-162.

Maritus, C.H.T. and Vlelc, P.L.G. (2001). The management of organic matter in tropical soils: what are the priorities? Nutrient cycling in Agro ecosystem. 61: 1-16.

Murwira, H.K. and Hirchman, H. (1993). Carbon and nitrogen mineralization of cattle manures subjected to different treatments in Zimbabwean and Swedish soils. In: Mulongoy, K.R. Merckx (eds.) soil organic matter sustainability of tropical agriculture. Pp. 189-198.

Nwoboshi, L.C. (1987). Tropical silviculture principles and techniques. Ibadan University Press. Pp.20-33.

Ogbe, F.M.; Egaregba, R.S.K. and Bamidele, J.F. (1992). Field survey of indigenous and useful plants. UN University Program. National Resourses in Africa pp. 124-126.

Ojeniyi, S.O. (2000). Effect of goat manure on soil nutrients and Okra yield in a rainforest area of Nigeria. Applied Tropical Agriculture 5: 20-23.

Okafor, J.C. (1981). Woody plants of nutritional importance in traditional farming system of the Nigerian humid tropics, Ph.D. Thesis, U.I. pp 179-185.

Oyenuga, V.A. (1967). Agriculture in Nigeria. An introduction No. 48 F.A.O. Rome, pp. 150-154.

Sara, W. (2005). Sustainable horticultural information, University of Sakatchewan, Extension Division, Department of Plant Sciences and Provisional Govrnment. Pp. 113-116.

Seifritz, W. (1982). Alternative and renewable source of energy in optimizing yields. The role of fertilizers. In: proceedings of the 12th congress. Pp. 153-163.

www.academicjournals.org.

www.hatworthpress.com/store/toc-Rss asp?

Research tables for the experiment

Table layout

Treatment	Quality		Replicate					
	Fi P/M	PL1	5	1	x	x	x	x
		PL2	10	2	x	x	x	x
		Pl 3	15	3	x	x	x	x
	F2 NPK	NPK1	5	4	x	x	x	x
		NPK2	10	5	x	x	x	x
		NPK3	15	6	x	x	x	x
	Control (c)			7	x	x	x	x

Table 1: Effect of Poultry Manure and N.P.K. Fertilizer on the Plant Height of *Chrysophyllum albidum* seedlings for Twelve Weeks

Average Plant Height (cm) of *Chrysophyllum albidum* Seedlings

Treatment	level (g)	2nd week	4th week	6th week	8th week	10th week	12th week	Mean—x
1	5	7.4	7.5	8.9	10.5	11.8	12.3	9.7
2	10	7.2	7.5	8.3	10.9	11.6	12.3	9.6
3	15	7.4	7.6	9.1	11.5	12.1	13.7	10.2
4	5	7.2	7.6	9.8	10.1	11.2	11.9	9.6
5	10	7.2	7.6	10.3	11.2	11.5	12.3	10.0
6	15	7.4	7.7	10.6	11.2	11.9	12.6	10.2
7	-	7.4	7.6	9.1	9.8	10.1	10.6	9.1

Table 1.1: ANOVA Table of the Treatment on Height of *Chrysohyllum albidum* Seedlings

SV	df	SS	MS	F-cal	F-tab
Block	5	145.31	29.06	104.91	
Treatment	6	5.91	0.985	3.56	2.42
Error	30	8.31	0.277		
Total	41	159.53			

LSD (0.05)

LSD = t(α/2,dfer) x √2MSE

R   LSD = 0.29039

Table 2: Average Stem Diameter (cm) of *Chrysophyllum albidum* Seedlings

Treatment	level (g)	2nd week	4th week	6th week	8th week	10th week	12th week	Mean— x
1	5	0.19	0.20	0.22	0.25	0.28	0.30	0.24
2	10	0.20	0.21	0.24	0.28	0.30	0.32	0.26
3	15	0.20	0.21	0.25	0.30	0.32	0.36	0.27
4	5	0.20	0.21	0.23	0.27	0.30	0.32	0.26
5	10	0.20	0.22	0.24	0.27	0.30	0.33	0.26
6	15	0.20	0.22	0.25	0.29	0.31	0.34	0.27
7	-	0.20	0.20	0.22	0.25	0.28	0.29	0.24

Table 2.2: ANOVA Table of the Treatment on Height of *Chrysohyllum albidum* Seedlings

SV	df	SS	MS	F-cal	F-tab
Block	5	0.0360	0.0072	16.56	
Treatment	6	0.00765	0.001275	2.33	2.42
Error	30	0.1305	0.000435		
Total	41	0.0567			

LSD (0.05)

LSD = t(α/2,dfer) x √2MSE

R   LSD = 0.00116g

7.6 Application Of Energy-Saving Ceramic Coolers Based On Evaporative Cooling In Hot Arid Regions.

Victor O. Aimiuwu[1]

Anthony U. Osagie[2]

Abstract

Evaporative cooling processes were known to some Ancient Civilizations. In Benin, it was used to provide cold potable water during the dry season, while the Persians used the methods to cool water and buildings in their hot arid environments. A physical model that explained these processes based on Newton's Law of cooling was developed by Aimiuwu in 1986. The temperature profile depends on several factors: the size of the pores in the ceramic materials, which varies with the firing temperature; an ellipsoidal shape to maximize the outer surface of the cooler; the convection currents around it, which quickly carry away water vapors from its vicinity, and the humidity of the atmosphere, which controls the rate of evaporation. Ceramic coolers have been used to produce cold water 15oC below the ambient temperature in three hours. In addition, the model predicts that the ceramic cooler can freeze its water content in three hours provided that 0.15% of the water content can evaporate every minute. The coolers can also be modified to preserve several common fresh fruits and vegetables for up to forty days. The following benefits are available to rural dwellers. Firstly, since the availability of electricity is not required, the operation of the ceramic cooler is not only possible but it also costs nothing. Secondly, rural farmers can easily and conveniently preserve their fresh fruits and vegetable without any fear of loss in quality between market days which may be four or more days apart.

Keywords: Evaporative Cooling, Ceramic Materials, Heat Exchange.

[1] Professor Victor Aimiuwu is Professor of Physics with the Department of Natural Sciences, Central State University, Wilberforce, OH 45385, USA

[2] *Professor Anthony Osagie is Professor of Biochemistry, Department of Biochemistry, Benson Idahosa University, Benin City, Nigeria*

1   Introduction

The prevailing high cost and environmental-unfriendliness of conventional energy sources are presently highlighting the importance of renewable energy sources and the need for energy conservation throughout the world. In the hot arid regions which are mostly located in developing countries, the climate is unusually very dry and hot as characterized by the following physical conditions: (i) very low annual rainfall and relative humidity; (ii) very high insulation and hot days; and (iii) rather fixed patterns of diurnal and seasonal winds. These conditions are favorable to solar, passive, and evaporative cooling processes [1-3], especially in rural areas where conventional electricity and its applications are neither affordable nor accessible.

2   Theory of Evaporative Cooling Using Fired Ceramic Materials

An evaporative cooling process takes place when molecules of water or any dissipating liquid gain enough energy to free themselves from the liquid state and go into the gaseous state without the net addition of energy to the liquid. These higher energy molecules leave behind a liquid with molecules of lower kinetic energy, which results in lower temperature. When such liquids are stored in porous containers, the cooling process is enhanced because more molecules with higher energy can travel through pores in the walls and escape from the outer surface of the container unlike in the case of the non-porous vessels. The ideal porous container should have sufficient porosity to allow enough seepage and evaporation without excessive liquid loss.

Aimiuwu fabricated such a ceramic cooler [4] as shown in Fig. 1, which has the following characteristics: (i) the container has an ellipsoidal shape in order to maximize its surface area and hence enhance the rate of evaporation; (ii) the materials is a clay mineral of the montimorillonite group, consisting of complex hydrous aluminum silicates ($Al_2O_3 \cdot 3\text{-}5SiO_2 \cdot nH_2O$) and usually includes Mg, Ca, and Fe; and (iii) the cooler is fired to 600 °C in order to maximize the pore sizes in the ceramic. This clay mineral has an irreversible phase transition temperature of about 570 °C when it changes from a plastic to brittle material. Firing to higher temperatures leads to a reduction of the pores by sintering processes [5].

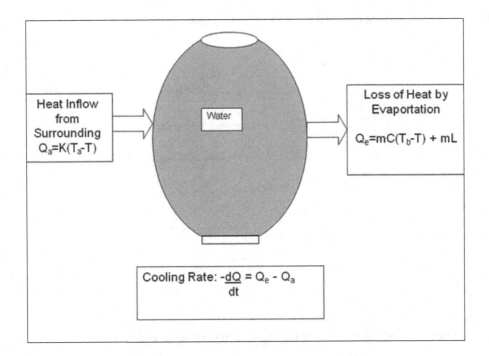

FIGURE 1: Shape and Energy Equations of the Ceramic Cooler.

The conservation of energy principle demands that the following equation must hold:

$$- (M + M_1\_mt)CdT/dt = mC(T_b\_T) + mL - K(T_a - T) \quad (1)$$

where M is the mass of water, M_1 is the water equivalent of the ceramic cooler, m is the mass of the water that evaporates per minute over its surface, C and L are the specific heat capacity and specific latent heat of water respectively, T_a is the ambient temperature, T_b is the boiling point of water and T is the instantaneous temperature of water in the cooler. It is shown elsewhere [4, 6] that the solution of eqn (1) is

$$T = (T_o\_T_e)e^{-\alpha\beta t} + T_e \quad (2)$$

where T_o and T_e are the initial and equilibrium (or final) temperature respectively of water in the cooler, α is the fraction of water that evaporates every minute over its surface while β is a heat exchange parameter between the cooler and its surrounding. Both α and β depend on several factors among which are (i) the humidity of the atmosphere, (ii) the convection currents around the cooler, (iii) the size of the pores in its walls, and (iv) its thickness. Table 1 and Fig. 2 show a modeling of eqn (2) for several values of α, β, and T_e. It is therefore theoretically possible to freeze the water in

Curve	a	b	c	d	e	f
T_e (°C)	25	20	15	10	5	0
β	122.13	61.07	40.71	30.53	24.43	20.36
α (10^{-4} s^{-1})	1.36	3.36	6.53	9.36	12.27	15.28

TABLE 1. Calculated values of α and β for the chosen values of T_e.

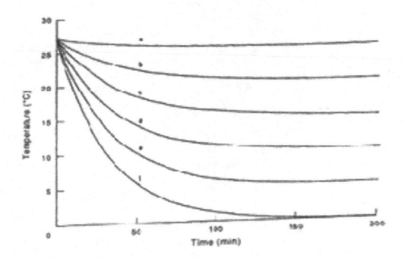

FIGURE 2. Theoretical Cooling Curves for Various Sets of α, β, and T_e.

The cooler as indicated in curve f, at no energy cost. The main condition that has to be met is that about 0.15 % of the mass of the water in the cooler must evaporate every minute all over its surface.

3   The Ceramic Cooler as A Cold Storage System

Two ceramic coolers, one fitting into the other, were fabricated from clay minerals as discussed previously [4, 6] and used to form a cold storage system. The inner one was a paraboloidal section while the outer cooler was an ellipsoidal section as shown in Fig. 3. The hourly variations of the temperatures of the inner cavity T and the surrounding T_a, as well as the difference between them, are shown in Fig. 4.

For a typical day, the hourly variations of the ambient temperature depend on several factors, such as insolation, wind, pollution, humidity, dust, etc., and are difficult to control. The daily temperature cycle is neither symmetric nor closed. It varied from a low of 15.7 °C at 05:00 h to a high of 33.5 °C at 14:00 h. The temperature of the inner cavity of the ceramic cooler varies from a low of 12.0 °C at 07:00 h to a high of 18.6 °C at 17:00 h. A comparison of the two curves reveals that: (i) the temperature of the ceramic cooler is consistently lower than that of the ambient temperature throughout the day; (ii) the daily fluctuation of 6.6 °C for the ceramic cooler is much less than that of 17.8 °C for the surrounding; (iii) the temperature cycle of the system lags behind that of the ambient by about 2 h; and (iv) the two cycles are neither symmetric nor closed. The most notable feature of the third plot is the maintenance by the device of a consistently high temperature difference of about 15 °C below the ambient temperature during the hottest part of the day [10].

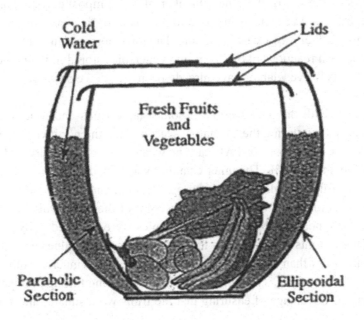

FIGURE 3. A Composite Ceramic Storage System

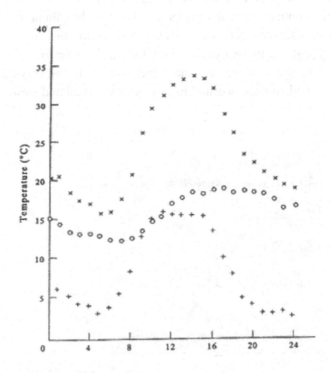

FIGURE 4. Hourly Temperatures of Ambient (x), Ceramic System (o) and Temperature difference between them (+)

These results suggested the use of the inner chamber of the composite system as a refrigerator for the storage of fresh fruits and vegetables in the rural provinces of the very hot and arid regions of the world where cold storage facilities are not available. The lower temperature of the ceramic cooler will slow down the breakdown reactions of the fruits and vegetables stored in it, and hence enhance their preservation. The results for vegetables have already been reported elsewhere [7, 8].

This presentation deals with the preservation of some staple fruits and root crops in West Africa. The research was carried out during the harmattan season when the north-east wind blows dusts and haze from the Sahara Desert across West Africa. The weather is generally dry and characterized by a relative humidity of less than 30 %. The inner chamber was wiped clean and dried. A known weight of a set of a particular fruit was placed inside an open cellophane sachet and then kept in the inner chamber and covered. Another set of almost equal weight of the same fruit was left exposed to the harsh weather on a table nearby. The two sets of samples were weighed daily for up to forty three days. The plots of the weights normalized to their respective initial values are shown in Fig. 5-7. A little amount of water (less than 5 cm^3) was always present in the inner chamber when the sets of fruits were removed for weighing. This collection of water, oozing through the pores of the walls of the inner cooler, has the advantage of ensuring that the fruits were situated in saturated water vapor at all times.

The results are very similar for the three cases. The samples exposed directly to the harsh weather lost weight very fast, leveling off at about 10 % of their initial weights in about two weeks. At the same time, the samples in the ceramic cooler gained weight slightly from 1.33 % to 7.09 % in the first few days before leveling off to remain fresh and preserved, The weight gain arises from a phenomenon known in botany as water saturation deficit (WSD) [9]. Before the fruits were harvested, they were losing moisture at a very fast rate to their dry environment. Consequently, the moisture content was less than what they would normally require to make them turgid. In the very humid chamber of the ceramic cooler, they absorbed enough moisture to make them normal and fresh.

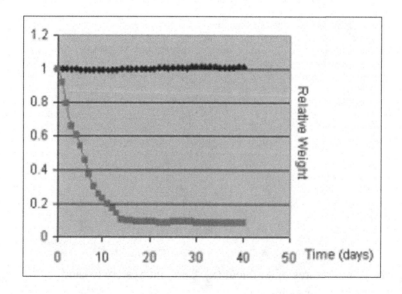

FIGURE 5. Preservation of Carrots (*Daucus carota*)

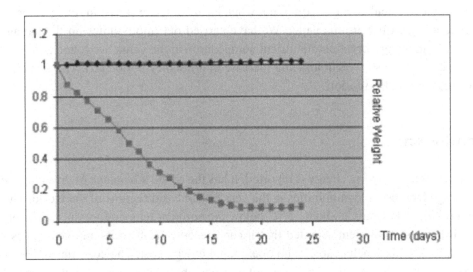

FIGURE 6. Preservation of Bell Pepper (*Capsicum frutescens*)

The exposed carrots (*Dalicus Carota*) lost weight steadily until there remains 8.48 % of its initial weight. The stored counterpart lost weight slightly (less than 0.86 %) during the first seven days before increasing its weight by 1.13 % eventually. It was preserved for forty days before rot set in. These data corresponds to WSD of 1.43 %. Similarly, the exposed bell pepper (*Capsicum Frutescens*) dried up steadily to 8.61 % of its initial weight in eighteen days and remained constant thereafter.

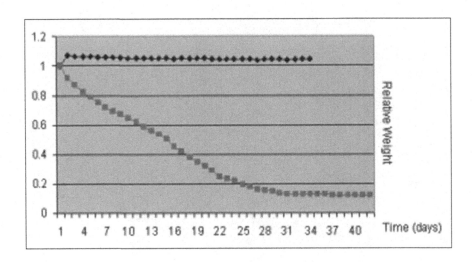

FIGURE 7. Preservation of Garden Eggs (*Solanum melongena*)

The stored set steadily gained 2.78 % in weight while remaining preserved for twenty five days. Bell pepper has WSD of 2.95 %. Lastly, the exposed garden eggs (*Solanum melongena*) dried up much more slowly to 12.04 % of its initial weight in forty days before leveling off. The stored sample gained 7.09 % of its initial weight by the end of the first day before reducing to 4.51 % by the thirty-fourth day. Garden eggs were preserved for thirty five days and has WSD of 7.46 %. It is desirable to compare the initial energy investment in firing the ceramic pots (usually on firewood burners) and the energy gain when cool water is harvested from the ceramic pots over several weeks.

It is also desirable to calculate the energy input when local fruits are stored in refrigerators as against the local practice described in this paper. We have carried out pilot studies on the storage of yam bulbils. Apart from the preservation of nutrient composition in the yams, we found a reduction in the magnitude and rate of tuber weight loss and increase in dormancy period (II). These studies deserve appropriate scale-up and extension.

4.   Conclusion

A cheap method for conserving energy is reported. It has the added advantage of providing cold water and cold storage facilities to communities in the very hot and arid regions of the world. The ceramic cooler can provide cold water at about 15 °C below the ambient temperature during the hottest part of the day. It has also been demonstrated that it can preserve fresh fruits and vegetables for up to forty three days. All this is achievable at no operational energy cost. The ceramic cooling method is regarded as crude by urban dwellers but it is not unacceptable. Adequate publicity is required if the method is to be generally adopted.

REFERENCES

1. Duffie J.A. and Beckman W.A. "*A Review of Solar Cooling*", Sharing the Sun **76(3)**, 13 (1976).

2. Bahadori M.N., "*Passive Cooling Systems in Hot Arid Regions*", Fifth Course on Solar Energy Conversion, 461 (1978)

3. Howe E.D., Sunworld **42**, 182 (1982).

4. Aimiuwu V.O., Nig. J. Appl. Sci. **6**, 65 (1988)

5. Budnikov P.P, "*The Technology of Ceramics and Refractories*", M.I.T. Press, Cambridge (1964).

6. Aimiuwu V.O., Energy Convers. Mgmt **33**, 69 (1992)

7. Aimiuwu V.O., Energy Convers. Mgmt **34**, 707 (1993)

8. Aimiuwu V.O and Ebeigbe S., African Tech. Forum **9(3)**, 1994

9. Eze J.M.O., Dumbroff E.B. and Thompson J.E., Physio. Plant **6**, 179 (1983)

10. Doogue E., Rolex Award for Enterptize, 24 (2003)

11. Osagie, A.U. The yam tuber in storage, Ambik Press, Benin City (1992)

7.7 Environmental Carcinogens And Sensitization Of Human Estrogen Receptor Positive Breast Carcinoma Cells To Paclitaxel And Vincrinstin By V. Amygdalina Extracts.

Carolyn B. Howard [1,3,4], Keyuna Cameron [2,3], Lecia Gresham [2,3] Yi Zhang [1,3], and Ernest Izevbigie [2,3,4]

This research was supported in part by the Extramural Research Centers in Minority Institutions (RCMI)/NIH grant # G122RR13459-07S1; Associates Research Development Award (EARDA)/ NIH grant #1G11HD046519-05; National Center for Minority Health Disparities (NCMHD)/NIH grant #P20MD000534-01

Phone: 601-979-3464; fax: 601-979-5853

E-mail: ernest.b.izevbigie@jsums.edu

Sabbatical Leave Address: Department of Basic and Applied Sciences, Faculty of Basic and Applied Sciences, Benson Idahosa University, P.M.B 1100, University Way, G.R. Benin City, Edo State, Nigeria.

---

[1] Breast Cancer Research Laboratory [2]The Laboratory of Cellular Signaling, Phytoceuticals, and Cancer Prevention and Therapies. [3]NIH-Center for Environmental Health, College of Science Engineering and Technology. [4]Department of Biology, Jackson State University, Jackson, MS 39217, USA.

Dr. Akanimo Odon and Dr. Sam Guobadia

Abstract

Breast cancer (BC) is the leading cause of death in women between ages 40 -55 and the second overall cause of death among women. The exact cause of breast cancer is still unknown but several known risk factors contribute to the onset of breast cancer. Besides genetic factors, petroleum products (hexane, benzene, jet fuel, toluene etc) and other environmental factors such as exposure to ionizing radiation, cigarette-smoking, and high-fat diets are risk factors for the development of BC. Several treatment options are available to women diagnosed with this disease including surgery, chemotherapy, radiation therapy and hormonal therapy. Paclitaxel (TAX) and Vincrinstine (VIN) are two chemotherapeutic agents used in the treatment of BC; however the side effects that accompany these drugs are numerous and undesirable. *Vernonia amygdalina* (VA), a plant native to Africa is consumed in the diets and used to treat several conditions. We have previously shown VA to have antiproliferative effects in cancerous cells. The objectives of these studies were to assess the ability of VA to synergize with known anti-cancer drugs (TAX and VIN) to inhibit the growth of Human Estrogen Receptor Positive (ER+) carcinoma cells. MCF-7 cell line, considered a suitable model to study ER+ breast cancerous cells. MCF-7 cells were propagated in RPMI-1640 medium, supplemented with 10% fetal bovine serum and 1% penicillin-streptomycin. Cell growth or inhibition was determined by DNA synthesis assays and confirmed by cell counts using a hemacytometer. Treatment of VA decreased DNA synthesis in a dose-dependent manner; 0.1, 0.5, and 1 mg/ml VA decreased DNA synthesis by 60, 65, and 80% respectively ($P<0.0001$). TAX treatment (10 or 100 nM) alone did not have any statistical effects DNA synthesis but 100 nM inhibited DNA synthesis ($P<0.05$) in the presence of 0.01 mg/m; additional of 0.1 mg/ml enhanced this effect ($P<0.002$) compared to control. Vincrinstine (VIN), neither 10 nor 100 nM alone. Consistent with the VA/ TAX interactions, 100 nM VIN treatment inhibited DNA synthesis ($P<0.05$) in the presence of as little as 0.01 mg/ml compared to control. These findings were further confirmed with mitosis assays using hematocytometer.

Key words: MCF-7 cells, Paclitaxel, Vincristine, *Vernonia amygdalina*, Breast cancer.

1   Introduction

Breast cancer (BC) is the most commonly diagnosed non-skin cancer and second leading cause of cancer-related deaths in women. Breast cancer represents 15% of new cases of all cancers. An estimated 182,460 women will be diagnosed with invasive BC and 40,480 women will die from the disease this year in the U.S. [1]. Postmenopausal women represent the highest incidence, although a significant number of younger women with hereditary predisposition are often afflicted with the disease. Length of life-time exposure to estrogens, ionizing radiation, cigarette smoking, genetic predisposition, high fat diet intake, and environmental factors such as petroleum products (hexane, benzene, jet fuel, toluene, etc) are the known risks factors associated with breast cancer [2-5]]. The agents of interest in this study are Paclitaxel (TAX) and Vincristine (VIN), two widely used chemotherapeutic breast cancer treatment drugs.

Paclitaxel (Taxol, TAX) is considered an effective anticancer agents used in the treatment of cancer [6]. It was isolated from the needles and bark of the Pacific yew tree (Taxus brevifolia). Taxol belongs to the taxoid class of antitumor compounds [7], which promotes formation and inhibits disassembly of stable microtubules, thus inhibiting mitosis. Paclitaxel displays a wide scope of antineoplastic activity, including effectiveness in treatment of metastatic BC [8]. The response rate of BC patients treated with TAX is 56% to 62% [9]. Treament with TAX results in numerous undesirable side effects, in addition to multidrug resistance (MDR). The side effects of TAX include, but not limited to, lowered resistance of infection, anemia, bruising or bleeding, tiredness and feeling weak, and diarrhea [6].

Vincristine (VIN) belongs to the cell cycle phase-specific Vinca alkaloid family. Primarily, the Vinca alkaloids induce cytotoxicity by tubulin interaction. There are many diverse biochemical and biologic actions unrelated to their effects on microtubules. These actions include competition for intracellular transport of amino acids, inhibition of purine, RNA, DNA and protein synthesis, disruption of lipid metabolism, and disruption in the integrity of the cell membrane and membrane function [10]. Vincristine has also been found to have antitumor properties by binding reversibly to spindle proteins in the S phase. Studies also indicate VIN as an inhibitor of RNA synthesis. Treatment with VIN induces neurotoxicity effects characterized by a periphereal, symmetric mixed sensory-motor, and autonomic polyneuropathy [10].

Plant-derived substances represent excellent sources for such novel and patient-friendlier anti-cancer and/or adjuvant agents. There is increasing evidence to show that *Vernonia amygdalina* (VA) may be one such agents. Vernonia amygdalina is a plant found in the African tropics. Leaves from this plant are consumed in the diet as a vegetable and as a culinary herb in soup in many African households [11]. Extracts of the plant is also used medicinally to control cough, constipation, fever, and hypertension [10-14]. Saponins, sesquiterpene, and flavonoids were isolated from a photochemical screening of VA by Igile et al 1994. [15]. VA has been shown to significantly decrease hypertriglyceridemia induced by dietary cholesterol [16]. It has been shown that VA used alone and in combination with chloroquine increased its efficacy in the treatment of malaria in vivo [17-18]. Njan et al., 2008 suggests that VA can be used as a palliative and alternative agent in the treatment of malaria [19]. Several mechanisms have been postulated which lead to the inhibition of cancerous cell growth by VA. In vitro and animal model studies suggest that VA, alone or in combination with known anti-cancer drugs, is inhibitory to the growth of several histogenic cancerous cells. Izevbigie et.al, reported VA as a potent inhibitor of mitogen-activated protein kinases in breast cancer cell lines [20].

In the present study, we hypothesized that VA will augment either the cytotoxic activities of TAX and VIN or sensitize the MCF-7 to TAX and VIN.

2   Materials And Methods

MCF-7 cells were purchased from American Type Culture Collection (ATCC, Manassas,VA). RPMI 1640 Medium, Fetal Bovine Serum (FBS), Antibiotic/Antimycotic and Phosphate Buffered Saline (PBS) were purchased from Fisher Scientific (Houston, TX). [³H]-thymidine (1mCi/ml) was purchased from MP Biomedicals (Solon, OH). All other chemicals were purchased from Sigma Chemical Company (St. Louis, MO. USA

Preparation of Aqueous Extracts of VA

Fresh pesticide-free VA leaves were collected in Benin City, Nigeria and rinsed with cold, distilled water. After rinsing, the leaves were spread out evenly on galvanized—wire screens with the edges bent upward 2 inches on all sides. The galvanized—wire screens were placed in a specially-constructed dryer and heated to 130-140°C for complete dryness within 4h. Three hundred grams of dried leaves were soaked in 6L of ddH$_2$O (1:20 w/w) overnight at 40°C before squeezing by hand to a mixture. The mixture was then filtered through a clean white gauge to remove the particulate matter before filtration through a 0.45-μm filtration unit for sterilization. The resulting sample solution was lyophilized to dry powder (30 g) on a SpeedVac Concentrator (Savant SC210A), transferred into a 50 mL centrifugation tube and stored at -20°C for bioactivity assays.

Cell Culture

Liquid nitrogen-frozen human breast adenocarcinoma (MCF-7) cells were thawed by gentle agitation of the vial for 2 min in a water bath at 37°C. After thawing, the content of each vial was transferred to a 75cm² flask, diluted with RPMI 1640 supplemented with 10% fetal bovine serum (FBS) and 1% penicillin-streptomycin, and incubated at 37°C in a 5% CO$_2$ incubator to confluence. The growth medium was changed every two to three days. Cells grown to 75-85% confluence were washed with phosphate buffered saline (PBS), trypsinized with 2 mL of 0.25% versene-trypsin-0.03% EDTA, diluted with fresh medium, and counted using a hemacytometer.

Cell Proliferation Determination

For determination of VA anti-proliferative effects alone and its synergistic effects with TAX, and VIN, the cells were seeded in100 mm tissue culture plates and allowed to grow to 60% confluence. Before treatment, the medium was aspirated and fresh medium was added. The cells were treated with 0.01, 0.1, 0.5, 1 mg/ml of VA alone. Concentrations of 0.01, and 0.1 mg/ml of VA was used in combination with 10, 100 nM Paclitaxel and Vincrinstine. The untreated cells were used as controls. After 24 h, medium was aspirated off adherent cells and the resulting monolayer was gently washed with 5 mL of PBS. The cells were collected by trypsinization and resuspended in RPMI 1640. The cell counts were carried out using a hemacytometer.

[³H] Thymidine Incorporation Studies

DNA synthesis was determined by [³H]thymidine incorporation assays. MCF-7 cells were seeded at a density of 5×10^4 in 35 mm diameter plates. MCF-7 cells were allowed to grow to 60% confluence before stimulating them with either VA, TAX, or VIN for 18 h. Treatments included 0.01 and 0.1 mg/ml of VA alone and in combination with either 10, 100 nM TAX or VIN. Twenty microliters (20 µl) of (1µCi/mL) of ³Hthymidine/35 mm well was added after 18 h incubation and incubated again for 6 h at 37°C. All incubations were terminated by aspirating the RPMI 1640 medium and doing triplicate washes with 2 mL of cold PBS to remove residual [³H]thymidine. Two milliliters (2mL) of 10% cold TCA was added to each well and incubated at 4°C for 10 min to fix the cells. Following fixation, the cells were washed three times with ddH$_2$O and solubilized by incubation for 30 minutes with 0.5M NaOH (2ml/35mm) at 37°C. Upon solubilization, 1 mL of cell solution was added to 5 mL of scintillation cocktail and mixed vigorously, and radioactivity was determined using a scintillation counter.

Statistical analysis

Triplicates within individual experiments were averaged and expressed as means ±SD. Comparisons between means were determined by unpaired students t-test with 2 tailed P values reported, employing GraphPad statistical software package. Data was determined to be statistically significant if P values were 0.05 or lower.

3   Results

Dose-dependence Inhibition of DNA Synthesis by VA Extracts

Inhibition of DNA synthesis is considered anticancer activity since DNA synthesis is required for cancer cell growth. VA extract inhibited DNA synthesis of MCF-7 cells in a concentration-dependent manner (0.1-1mg/mL). Treatment of VA decreased DNA synthesis in a dose-dependent manner; 0.1, 0.5, and 1 mg/ml VA decreased DNA synthesis by 60, 65, and 80% respectively (P<0.0001) (Fig. 1)

Cotreatment of cells with VA extracts and TAX

Treatment of cells with 0.1 mg/ml VA extracts inhibited DNA synthesis approximately three fold compared to control (P < 0.0026). Neither 10 nor 100 nM alone TAX treatment of cells had significant effect on DNA synthesis, but 100 nM TAX significantly decreased DNA synthesis (P<0.05) in the presence of as little as 0.01 mg/mL VA extracts (Fig. 2).

DNA Synthesis Inhibition by different concentrations of VA and Vincristine

Exposure of cells to either 10 nM or 100 nM VIN alone had no effects on DNA synthesis, but 100 nM VIN significantly decreased DNA synthesis (P < 0.05) in the presence of 0.1 mg of VA extracts compared to control (Fig. 3). These findings are consistent with the DNA synthesis inhibitory action of VA and TAX (Fig. 2).

Cytotoxic Activities of VA Extracts, TAX and Combination of VA Extracts and TAX

Vernonia amygdalina (0.1 mg/ml) elicited a modest cytotoxic activity (decreased cell viability) compared to untreated controls (Fig.4) which was inconsistent with the 3-fold inhibitory effect on DNA synthesis observed in the DNA Synthesis assays (Fig 1). This may be explained, at least in part, that IC_{50} for DNA synthesis is much lower than for cytotoxicity [21]. Taxol (TAX) at 10 nM had no significant activity on cell viability to controls (Fig. 4) but, 100 nM TAX inhibited cell viability ($P < 0.0004$) and the effect was enhanced by the presence of 0.1 mg/ml VA extracts (Fig. 4).

Cytotoxic Activities of VA and VIN

Like VA extracts, Vincrinstine treatment (10 nM and 100 nM) decreased cell viability whereas, 10 nM VIN treatment did not elicit significant effects on cell viability but decreased cell viability $P<0.05$) in the presence of as little as 0.01 mg/ml VA compared to control (Fig 5). The DNA Synthesis Assays were confirmed by the Cytotoxicity Assays.

4   Discussion

The conventional therapies (Paclitaxel and Vincristine) used in the treatment of cancer cause numerous side effects. These side effects (including a low white blood cell count, weakness, infection, and muscle pain, as well as numbness, tingling, neuropathy, cardiac ischemia, and alopecia) limit the therapeutic values or benefits of the drugs. [22-23]. Considering the impediment the un-wanted side effects pose to treatment effectiveness, the search for novel agents that ameliorate and/or improve the activities of existing drug is timely. *Vernonia amygdalina* leaves and extracts have been used in African cultures as a vegetable in the diets and as a medicinal alternative for several common ailments without any adverse side effects [24]. In the present study, we have shown that exposure of cells to 0.1, 0.5, and 1.0 mg/ml VA inhibited cell growth by 60, 65, and 85% respectively (Fig. 1) which is consistent with previous reports [20,21,25). Neither TAX nor VIN at concentrations used (10 and 100 nM) alone had significant effects on DNA synthesis but 100 nM of either TAX or VIN significantly inhibited DNA synthesis ($P<0.05$) in the presence of as little as 0.01 mg/ml of VA extracts (Fig. 2 and 3). Studies have shown VA as having anticancer properties [20,21,25]. The present study validates these findings. In conclusion, the present findings accomplished the following: corroborate previous reports that VA extracts inhibits the proliferative activities of cancerous cells; and that VA may sensitize ER+ human breast carcinoma cells to chemotherapy. In the Cytotoxicity Assays, VA synergized with TAX (100 nM) thus increasing the cytotoxicity of TAX (Fig. 4) while VA also sensitize cells to 100 nM VIN treatment. Taken together, the present study present evidence that VA may posses cell-sensitization to chemotherapy and chemotherapy-augmentation activities. Therefore VA may be used as adjuvant with TAX or VIN to improve treatment outcome.

Acknowledgements

This research was supported in part by the Extramural Research Centers in Minority Institutions (RCMI)/NIH grant # G122RR13459-07S1; Associates Research Development Award (EARDA)/ NIH grant #1G11HD046519-05; National Center for Minority Health Disparities (NCMHD)/NIH grant #P20MD000534-01.

REFERENCES

American Cancer Society, 2009

Schultz Wolfgang Arthur: Molecular Biology of Human Cancers: An Advanced Student's Textbook

Gun RT, Pratt NL, Griffith EC, Adams CG, Blisby JA, Robbinson KL. Upgrade of a propective study of mortality and cancer incidence in Australian Petroleum Industry. Occup Environ Med. 6(12): 150-156, 2004

LoPresti E, Sperati A, Rapiti E, Domenicantonio R, Forastiere F, Perucci CA, Med Lav 5: 327-337, 2001

Rushton L, Alderson MR. A case-control study to investigate the association between exposure to benzene and deaths from leukaemia in oil refinery. Br J Cancer 43(1): 77-84, 1981

Miller ML, Ojima I. Chemistry and Chemical Biology of Taxanes Anticancer Agents. 1(3):195-211, 2001

Huizing MT, Misser VH, Pieters RC, ten Bokkel Huinink WW, Veenhof CH, Vermorken JB, Pinedo HM, Beijnen JH. Taxanes: A New Class of Antitumor Agents. Cancer Invest. 13(4):381-404, 1995

Pharmacological Actions of Paclitaxel. 6(31):27-9, 1999

Rowinsky EK, Onetto N, Canetta RM, Arbuck SG. Cancer Therapy and Biotherapy.

Chabner Bruce, Longo Dan Louis. Cancer chemotherapy and biotherapy: principles and practice Edition: 4, illustrated Published by Lippincott Williams & Wilkins, 2005

Argheore EM, Makkar HPS, Becker K. Feed value of some browse plants from central zone of Delta State of Nigeria. *Trop Sci.* 38:97-104, 1998

Regassa A. The use of herbal preparation for tick control in Western Ethiopia. *J S Afr Vet Assoc.* 2000;71:240-3.

Amira CA, Okubadejo NU. Frequency of complementary and alternative medicine utilization in hypertensive patients attending an urban tertiary care centres in Nigeria. *BMC Complementary and Alternative Medicine.* 7:30-48, 2007

Kambizi L, Afolayan AJ. An ethnobotanical study of plants used for the treatment of sexually transmitted diseases (njovhera) in Guruve District, Zimbabwe. *J Ethnopharmacol.* 77:5-9,2001

Igile GO, Oleszek W, Jurzysta M, et al. Flavonoids from *Vernonia amygdalina* and their antioxidant activities. *J Agric Food Chem.* 42:2445-8, 1994

Oluwatosin A Adaramoye, Olajumoke Akintayo, Jonah Achem, and Michael A Fafunso. Lipid-lowering effects of methanolic extract of *Vernonia amygdalina* leaves in rats fed on high cholesterol diet. Health Risk Manag. 4(1): 235-241, 2008

Iwalakon BA. Enhanced antimalarial effects of chloroquine by aqueous Vernonia amygdalina leaf extract in mice infected with chloroquine resistant and sensitive Plasmodium berghei strains. Afr Health Sci. 8(1):25-35, 2008

Abosi AO, Raseroka BH. In vivo antimalarial activity of Vernonia amygdalina. Br J Biomed Sci. 2003;60(2):89-91.

Njan AA, Adzu B, Agaba AG, Byarugaba D, Diaz-Llera S, and Bangsberg DR. The analgesic and antiplasmodial activities and toxicology of Vernonia amygdalina. J Med Food. 11(3):574-81, 2008

Izevbigie EB, Bryant JL, Walker. A novel natural inhibitor of extracellular signal-regulated kinases and human breast cancer cell growth. Exp Biol Med 229(2):163-9, 2004

Opata M, Izevbigie EB. Aqueous Vernonia amygdalina Extracts Alter MCF-7 cell member permeability and efflux. Int. J. Environ. Res. Public Health, 3(2): 174-179, 2006

National Cancer Institute. **New Form of Paclitaxel Causes Fewer Side Effects in Advanced Breast Cancer.** www.cancer.gov.

Pratt WB. The anticancer drugs. Oxford University Press. US. 1994.

Adaramoye O, Ogungbenro B, Anyaegou O, Fafunso M,. Protective effects of extracts of vernonia amygdalina, hisbicus, sabdaiffa, and vitamin C against radiation-induced liver damage in rats. J. Radiat. Res. 49: 123-131, 2008

Izevbigie EB. Discovery of water-soluble anticancer agents (edotides) from a vegetable found in Benin City, Nigeria. Exp Biol Med. 228(3):293-8, 2003

List of Figures

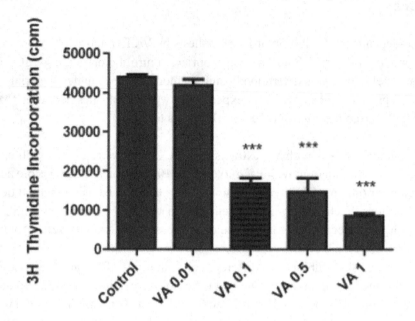

FIGURE 1

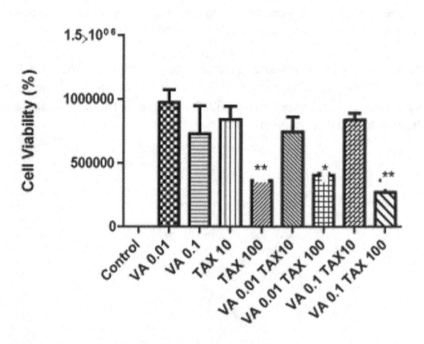

FIGURE 2

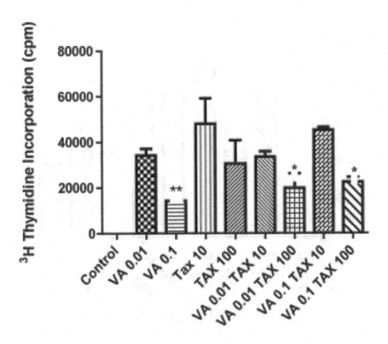

Concentrations VA (mg/ml); Paclitaxel (nM)

FIGURE 3

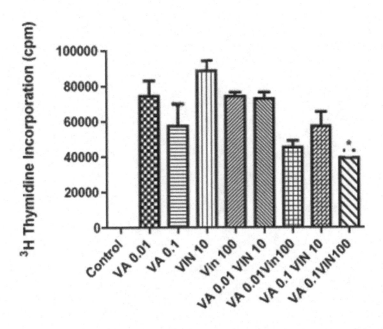

Concentration of VA (mg/mL); Vincristine (nM)

FIGURE 4

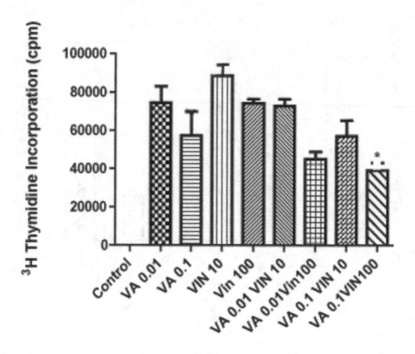

Concentration of VA (mg/mL); Vincristine (nM)

FIGURE 5

7.8 Building Intellectual Capacity For Academic Development In Nigeria: A Research Perspective

Oluwole, E.A., Attahiru, H., And Oyediran, O.B.

Abstract

Research in Nigeria has been suffering from initiative fatigue and academic development process has weakened its sense of progress and the ability to convince others that research development has all it takes at its core to move the nation forward. This paper examines the need to build intellectual capacity for academic research development, as a result of the growing concern about the future of scientific research, the next generation of researchers, and the dynamism of research needs and procedures in our country. Building research capacity in the nation is important to the growth of the economy as it helps to facilitate the acquisition of information and knowledge on pertinent research issue, provide training to enhance skills and assists to combat the numerous challenges facing research development. To achieve these, we have identified building intellectual capacity to acquire and use existing knowledge and at the same time building capacity to produce and use new knowledge champion research development among other alternative.

Keywords: Research development, Academic development and Capacity building.

1   Introduction

Within the realm of tertiary institutions many things are always changing; the structure of the educational system, curriculum, mode of teaching, methods of manpower training, the amount and types of provisions for institutions, such as laboratories, libraries, research instruments, lecture hall supplies, furniture and so on. These changes may lead to an improvement, or a worsening in the quality of an educational system (UNESCO, 2005). Academic research development usually encompasses discovery, innovation, experimentation, and creation; however in today's highly competitive and global economy it also involves technological transfer and partnerships with industry. In other words, academic research is about creating new knowledge and building new bridges while having a positive impact on the community and the country in general.

Research is an organized study or a methodical investigation into a subject in order to discover facts, to establish a theory or to develop a plan of action based on the facts discovered. It is undertaking systematic investigation in order to discover new facts or get additional information to elucidate a particular problem, and arriving at dependable solutions to problems, through planned and systematic collection, analysis and presentation of data. (Encarta, 2009, Rahman, 2005 and Osuala, 1993). Broadly, researches have been classified into two types: pure and applied, depending on the approach and objectives it set to achieve. Pure research is to expand the frontiers of knowledge, which involves conscious and persistent efforts to discover something new to enrich human knowledge in a fundamental fashion. While applied research is conducted to find answers to scientific problems, test facts and propositions of basic or pure research. (Okpamen, 2000).

The concept of academic research is a broad one. Many people see academic research from different perspectives, depending on their role in the academic system. The lecturers, researchers, policy makers, investors, parents or students would consider the subject from his own point of view and probably in terms of what use he would make of it. For example, Court, (2004) puts it that from individual researcher's perspective we investigate topics about which we are curious or passionate as well, we do research because it is an integral part of the academic role and a central factor in academic promotion.

Considering that academic research could address a multitude of issues and problems in the academia and the country at large. And that its finding could go beyond the researcher to different groups of people involves in policy reform and new product development, it becomes obvious it can lend itself a definition. Deighton (1971) defined academic research as those activities which are initiated within or based directly on the findings and methodologies of the social, behavioural and information science, and which are oriented towards the improvement of education and instruction. Harnquist (1973) saw academic research as all kinds of 'disciplined inquiry' in the academia. In his words academic research not only comprises empirical enquiry with quantitative techniques but also naturalistic observation, logical and philosophical analyses and historical. It can take place within the conceptual and methodological framework of different academic discipline such as psychology, sociology, economics, philosophy and history.

Thus enhancing the capacity of research in our universities, research institutions or agencies to carryout academic research in line with their mandate and discipline provide an interesting and potential spring board for research development in Nigeria. The United Nation Development Program (2010), defined capacity as the ability of individuals, institutions and societies to perform functions, solve problems, set and achieve objectives in a sustainable manner. Capacity consists of two major aspects: human and institutional. Human capacity is the stock of trained, skilled and productive persons capable of performing the essential tasks necessary for an organization to realize its goals or developmental agenda. While institutional capacity refers to the available organizational structures and processes which facilitate the achievement of developmental objectives. (Itsede, 2003).

Building intellectual capacity for academic research is not only to undertake research projects, but it's about the ability to provide efficient and sufficient research capacity in terms of materials and human or intellectual resources, that will engage in the whole process of research. That is building research capacity consists of strategies, tools, processes, procedures and structures that are employed in a given research or academic institution aimed at improving the capacity of its workforce to undertake educational research. From the foregoing it is evident that the growth, development and success of academic research are anchored on a more robust human and institutional capacity profile.

2   Importance Of Capacity Building To Academic Research

Capacity building is an essential lubricant for our development, more important that finance. Capacity building provides a country and its people with the ability to design, implement and monitor sound macroeconomic and sectorial policies for efficient management. This helps to layout the essential national and institutional prerequisite for economic and social progress as it assist in need assessment and also helps individuals to realize their potentials for a better career, and enables indigenous ownership of developmental agenda through academic research.

The need for intellectual capacity building in research is eminent as approaches to academic development are changing, over the years from promoting the right ethics to filling the gap in skills and service provision. These needs can be strengthened across the nation by increasing the awareness about the strategic importance of research, by creating an enabling environment where researchers and academic institutions can take advantage of resources available to fulfil their mandate in order to have a meaningful role in the development process of the nation.

The overall objectives of building intellectual capacity are:

* to facilitate the acquisition of intellectual information and knowledge on pertinent research issues;
* to provide awareness or training in the use, evaluation and management of material resources for research;
* to enhance the skills of researchers in the preparation, production and management of journals and;
* to reverse the brain-drain syndrome bedevilling tertiary institutions in Nigeria and engage the skills and energy in the diasporas.

The above mentioned essentials of capacity building are geared towards academic research development through access to knowledge and skills, innovation and proven methodologies, networking and funding opportunities which are viable options for institutional management and governance.

3   Tertiary Institutions And Academic Research Development

Education is widely accepted as an instrument for promoting socio-economic, political and cultural development in Nigeria. Tertiary institutions like universities and colleges—educate future leaders and develop the high-level technical capabilities that underpin economic growth and development (Odekunle, 2001). The main purpose of tertiary education in Nigeria is to produce the much needed manpower through research to accelerate the socio-economic development of the nation. Because such specialized academic research at high level are regarded as instrument of social change and economic development. However, despite the immense benefit of academic research to nation building, yet the potentials of tertiary institutions and indeed research institutions in Nigeria to fulfil their responsibility is frequently jeopardized by the long standing problems bedevilling the sectors. Such as inadequate funding, discrepancy in the allocation of research resources, issue of academic freedom and autonomy, reoccurring strike action by unions, cultism and student unrest among others.

Gosling, (2009) believes that academic development deals with making decision about the scope of 'development' based on pragmatic consideration within a context dependent framework, determined by what is possible within that particular circumstance, including key factors such as the size and culture/mission of the institution, the level of priority given to teaching/research and the size and capacity of its workforce. In other words, academic development is a field of activity which has no strong boundaries, is unstable and is necessarily required to respond to the environment, in which it operates.

A look at the 'concept' of academic research development, suggest that there are three elements to it. First what kind of academic research are being develop? Secondly to what end are they being developed? And thirdly, what is the process through which the subject is being developed in order to achieve the desired end? The early academic developers were driven by a strong desire to improve teaching, to make things better, according to their view of what would make a better experience for the academia, and this was the goal they achieved. That is why the National Policy on Education (2004) saw tertiary institutions as canters of excellence, teaching, research and storehouse of knowledge, which are expected to:

1. Contribute to national development through high-level relevant manpower training.
2. Develop the intellectual capability of individual to understand and appreciate their local and international environment.
3.   Acquire both physical and intellectual skills, which will enable individuals to develop into useful members of the community.
4. Develop and inculcate proper value orientation for the survival of the individual and the society.

Nakoda (2009) believed that putting forward some well-articulated aims does not bring success to any organization unless such aims are married with good management and effective teaching and learning atmosphere. Land (2000) also pointed out that through rational means an academic institution can be improved-as change is seen as progress, and change process can only be understood through scientific principles or 'evidence based practice'. But Taylor (1973) concluded that the necessary conditions for effective research development are adequate resources, appropriate structure and a sympathetic political, social and educational climate.

The roles of academic research are wider than mere relation of research practice as confirmed by (Dockrell, 1980). He considered the roles of academic research to include:

i. The traditional provision of research findings which could make suggestions for policy rather than prescribe authoritative guidance.
ii. The stimulation of thought by drawing new issues, questioning assumptions and pointing out weakness of current practice and;
iii. Contribution of form and structure to the evaluation of new curricula and techniques.

Research of this kind aimed at increasing the problem solving capacity of the academic system, rather than to provide final answers to questions or objective evidence to settle controversies (Nisbat, 1980). It became obvious that although academic research may be pursuit for knowledge, but much of the outcomes of such research do influence educational practices and the economy in many ways.

4   Building Intellcetual Capacity For Research Development

There appears to be an emerging consensus that capacity building is an essential tool for sustainable development; and haven identified the need to build intellectual capacity research as a precondition for sound academic development, the questions on answered are 'how do we build this capacity? And how should policy makers allocate scarce resources to this capacity building objective? Generally, research capacity for academic development could be build from two points which are specifically for education and training institutions, and research institutions.

1. Building intellectual capacity to acquire and use existing knowledge.
2. Building intellectual capacity to produce and use new knowledge.

To acquire and use existing knowledge simply means to import, adapt and adopt knowledge produced outside the country. According to (Mathews, 2002) the most critical aspect of the catching up process is the absorption, adoption, and adaptation of products, processes, and technologies that are already in use elsewhere. Acquiring such knowledge that was developed outside the country, adapt it for local use, diffuse it throughout the nation, and adopting it locally—is a major conduit for building capacity in every country irrespective of its level of development. Even if a country dramatically increases the size and quality of its research efforts, it is unlikely that the local research and development systems will generate more than a small fraction of the total knowledge needed by the country, (World Bank, 2008).

Diffusion is not a passive process as observed be Mathews (2002). It is not something that simply happens to an enterprise or an economy. It requires an active conscious policy of linkage, leverage and learning. Nigeria has a distinct advantage if we can skilfully recognize and develop tools and strategies for exploiting existing research knowledge rather than devoting time, resources, and effort to develop new research model from the scratch. Just as Juma (2006) argued that Africa potentially has access to more scientific and technological knowledge than the more advance countries had in their early stages of industrialization. Thus the task for Nigeria will be to develop the research skills required to make use of this knowledge, such a way that the entire process of research development will be buttressed, supported and disciplined by an institutional framework that will ensure success and sustainability. Hence, most of the knowledge that our country will need if it is to grow and prosper will be produced by others.

The World Bank believe that the capacity to produce and use new knowledge via research may entail the capacity to conduct high-level basic research alone, or in partnership with leading global research and development institutions. Or it may entail building the capacity fine novel ways of solving local problems. Not every country has the current capacity to (or pressing need) participate in the global research and development effort to find a cure for HIV/AIDs or to develop an anti-malarial vaccine. But every country needs to develop the research and development capacity needed to find new, innovative ways to apply modern science to solving local problems. This may be difficult to achieve, but indigenous research institutions often contributes to the process based on local knowledge.

Tertiary and research institutes could be the main transmission mechanism between the global stock of research knowledge and intellectual development. As more skilled and equipped researchers are prerequisite for producing more knowledge in research development. This is because the availability of skilled and equipped researchers in our institutes translates to more knowledge advancement in research. But due to the fact that our research institutions operate in isolation from each other, and more importantly research institutes and universities in Nigeria do not collaborate with each other or work closely; research has being primarily conducted in independent laboratories and institutes that frequently set priorities without regard for market demand, technology up-grading, competition or government's own scientific priorities.

Clarke and Ramsey, (1990) pointed out that for these institutions to be successful, both in their studies and in the world of work, another prerequisite is a need to bridge the all-too-frequent gulfs between educational research and education or professional practices. For example, experience suggests that modern research functions bests when:

(i) Research is linked to teaching;
(ii) Scientists and engineers from different discipline collaborate in multidisciplinary problem solving teams, rather than working alone;
(iii) The supposed distinction between basic and applied research are minimized or eliminated; and
(iv) There is close linkages between researchers and business enterprises.

Thus building such capacity involves infrastructural investment by providing the physical facilities needed to tap into the existing poll of global knowledge. Because spending more on research facilities may not results in the desired economic benefit unless additional resources are accompanied by the necessary organizational and structural changes. Even Nigeria, with limited existing research capacity and budgets will need to decide whether to focus on cutting-edge research or on research designed to support the economy's capacity to import and adapt existing technology. In addition, we will need to decide what capacity can be built internally within the country and what capacity needs to be built on regional basis with other countries.

Challenges Of Academic Research Development

Adeyinka (1975) affirms that Just as the physical development and social development of the average child is beset with many problems, so is the development of academic research in any given society is hampered by a variety of problems,—retarding the pace of educational development in Nigeria today. Despite the critical role of academic research it has suffered a lot of setbacks, in the country. Ajayi and Ayodele, (2004) argued that higher education in Nigeria is in travail, the system is riddled with crises of various dimensions and magnitude. A number of multi-faceted problems, have inhibited goal attainment and are raising questions, doubts and fears, all of which combine to suggest that the system is at a cross road.

One major problem is the overall lack of funding and discrepancy in the allocation of research funds as stated by (Rahman, 2005). Ajayi and Ayodele, (2004) also confirmed that there was no increase in the proportion of total expenditure devoted to education, but this has been considered to be rather grossly inadequate considering the increasing cost of administration, which has been aggravated by inflation. The Nigerian government over the years has not been meeting the United Nations Educational Scientific and Cultural Organization (UNESCO) recommendation of 26% of the total budget allocation to the educational sector as submitted by Ajayi and Ekundayo (2006). This growing shortage of funds has apparently been responsible for the decline in learning and research resources in academic institutions in Nigeria.

Currently the organization of research seems to violate the ideal precept particularly when the boundaries between applied and basic research are becoming increasingly blurred, different ministries may be responsible for basic research and for applied research. Teaching and research now take place in separate institutions, with little or no interaction between the two. Also are organized vertically, with physicists in one institute, mathematicians in another, and chemists in yet another institute, rather than in broader, multidisciplinary problem-solving teams. Most research institutions frequently operate in isolation from each other, and even worse research institutes and universities do not collaborate with each other or work closely and research are conducted in independent laboratories and institutes without regard for market demand, technological up-grading, competition or government developmental priorities.

The state of infrastructure, facilities and equipment required for academic development are in shambles. The World Bank in 2004 reported the equipment for teaching and research and learning are either lacking or very inadequate and in bad shape to permit (researchers) the freedom to carry out their basic functions of academics. The World Health Organization (WHO) cited by Rahman, in 2005 attributed poor research output to lack of sophisticated equipment and trained manpower to maintain such equipment. All the resources required for academic development are in short supply; lecture halls, laboratories, student's hostels, library-spaces, books and journals, office spaces are seriously inadequate, (Ochuba, 2001).

Another challenge to academic research development is the issue of academic freedom and autonomy in tertiary institution in Nigeria. According to Babalola et al, (2007) university autonomy and academic freedom has over the years been a recurring issue in ASUU's demand from the federal government. Academic freedom and autonomy has been described in the literature as protection of tertiary institutions from interference by government officials in the day today running of the institutions and in the appointment of key office holders.

Akindutire (2004) submitted that institutional deterioration and salary erosion during the past decade have prompted substantial 'brain-drain' of academic staff and impeded new staff recruitment. Brain-drain refers to widespread migration of academic staff from tertiary institution in the country to overseas universities or equivalent institutions where their services are better rewarded. It must be emphasized that while the best brains are leaving the country, the broad aim of producing high level manpower, through research for national development cannot be achieved.

5   Recommendations and Conclusion

When we look at the divisive issues of academic research development, there is little or no solace in the current state of research in Nigeria. Although tertiary institutions in the country have produced abundant skilled manpower in the area of research, who are found in the universities, research institutes and other government ministries, but the research activities in these three milieus are different and intellectual capacity for is insufficient. Research in our universities are mostly small individual projects undertaken for promotions; which conflicts with government's need for research that would contribute more directly to developmental objectives. There is also little consideration for research in terms of priorities or identification of basic research problems. Research in Nigeria not only lacks funding, but also the sensitivity to frequent changes in technological and political climates.

What bench-mark of sound academic development have we, as academicians and researchers offered as alternatives? What coherent and comprehensive picture of growth and development have we placed before the public as a reasonable counter or even counterproductive to our research experience? Because, currently there is no feasible alternative in terms of comprehensive model and theories for academic development as pointed out by Alexander (2002). The academias still have to embrace the concept of research development and consequently, the research communities also have to immerse them self in all kinds of longitudinal, multi-dimensional, multi-method studies that would articulate credible framework for research.

REFERENCES

Adeyinka, A.A. (1975). Current Problems of Educational Development in Nigeria. *The journal of Negro Education*. Vol.44, No.2. Pp 177-183.

Ajayi, I.A. and Ayodele, J.B. (2004). Fundamentals of Educational Management. Ado Ekiti: Green-Line Publishers.

Ajayi, I.A. and Ekundayo, H.T. (2006). Funding Initiative in University Education in Nigeria. A paper presented at the National Conference of National Association for Educational Administration and Planning. Enugu State University of Science and Technology.

Akindutire, I.O. (2004). Administration of Higher Education, Lagos: Sunray press.

Alexander, P.A. (2002). "Toward a Model for Academic Development: Schooling and acquisition of knowledge". *Educational Researcher*. Vol.29. No.2. Pp 28-33.

Babalola, J.B., Jaiyeoba, A.O., and Okediran, A. (2007). University Autonomy and Financial Reforms in Nigeria: Historical background, issues and recommendation from experience. In J.B. Babalola and B.O. Emunemu (eds). Issues in Higher Education: research evidence from sub-Saharan, Africa. Lagos: Bolabay Publications.

Clarke, E. and Ramsey, w. (1990). "Problem of Retention in Tertiary Education" *Education Research and Prospective*. University of West Australia. Vol.17. No.2. pp 47-59.

Court, D. (2004). "The Quest for Meaning in Educational Research." *Academic Exchange Quarterly*. Volume 8, Issue 3.

Deighton, L.C. (1971). *The Encyclopaedia of Education*. Vol. 7, New York: The Macmillan Co and the Free Press.

Encarta, (2009). "Research and Development" Microsoft Encarta 2009 (DVD). Redmond, W A. Microsoft Corporation, 2008.

Federal Republic of Nigeria (2004). The National Policy on Education. Lagos, NERDC.

Harnquist, K. (1973). "The Training and Career Structure of Educational Research" in W. Taylor (ed), Research Perspective in Education. London: Rout-ledge and Kegan Paul Ltd.

Gosling, D. (2009). "Academic Development Identity and Positionality." Available on www.

Itsede, O.C. (2003). "The Role of Human Resource Development in NEPAD Agenda". *CBN Economic and Financial Review*. Vol. 41, No. 4 pp.107-120.

Juma, C. (2006). "The New Culture of Innovation: Africa in the Age of Technological Opportunities" Keynote Paper presented at the 8th summit of the Africa Union, Addis Ababa, Ethiopia.

Land, R. (2000). "Orientation to Educational Development" Educational Development pp 19-23.

Mathews, J.A. (2002) "Competitive Advantages of the Latecomer Firm: A Resource-Based Account of Industrial Catch-Up strategies. *Asia Pacific Journal of Management*. Vol. 19. No. 4, pp467-488.

Nakpodia, E.D. (2009). "Implications and Challenges of Nigeria University As Learning Organizations. *Academic Leadership Journal*. Vol. 7, Issue 3. Retrieved from www.academicleadership.org. Accessed 20/08/2010.

Nisbet, J. (1980). Educational Research: 'The state of Art' in W.B. Dockerell and D. Hamilton (eds). Rethinking Educational Research. London: Holder and Stoughton.

Ochuba, V.O. (2001). Strategies for Improving the Quantity of Education in Nigerian Universities. In N.A. Nwagwu, E.T. Ehiametalor, M.A. Ogunu and M. Nwadiani (eds). Current Issues in Educational Management in Nigeria. A publication of the Nigerian Association for Educational Administration and Planning.

Odekunle, K.S. (2001). "Funding of University Education under Democratic Rule in Nigeria: Problems and Prospects". Proceeding of the 12th General Assembly of SSAN.

Okpamen, P.E. (2000). Social Science Research Methods. Okasun concept Ventures. P8.

Osuala, E. C. (1993). Introduction to Research Methodology. Benin City: Ilupeju Press.

Rahmam, G.A. (2005). "Scientific Medical Research and Publication in Nigeria". *Nigerian Journal of Surgical research*. Vol.7, No. 3-4. Pp244-250.

Taylor, W.L. (1973). "The Organization of Educational Research in the United kingdom". In W. Taylor (ed). Research Perspectives in Education. London Rout-ledge and Kegan Paul Ltd.

UNDP, (Update,2010). Capacity Building for International Development. Available on www.oneworld.net/capacitybuilding.html. Accessed 12/03/2010.

UNESCO, (2005). Educational research: Some basic Concepts and Terminology.

World Bank, (2008). Science, Technology and Innovation: Capacity Building for Sustainable growth and Poverty Reduction. Washington D.C.

7.9 The Balance Sheet Of Agricultural Production In The Past Years In Nigeria

Dafe K. G. Ayanru

Abstract

Perhaps for the first time in Nigeria's history we had in place a civilian or democratic government (the General Olusegun Obasanjo's administration) whose avowed number one priority was agriculture. It is therefore a unique opportunity to examine the records at this time to see if the good intensions were met with needed action, as this has not always been the case in the past. This paper presents data to show the following, among others: i) that the consequences of Nigeria's declining agricultural fortunes have been glaring and far-reaching; ii) that African agriculture—especially Sub-Sahara Africa of which Nigeria is part—faces a series of chronic, acute and interacting barriers; iii) that political opinion varies as to the causes of Nigeria's agrarian retrogression; iv) that since the 70s government has initiated a series of impressive, high sounding but ineffective agricultural programmes; v) that the average national demand and supply for the main raw materials in Nigeria reveal a glaring deficiency in the agricultural sector; vi) that estimated hecterages and submarginal yields of some major crops in Nigeria, which show a downward trends, give further cause for concern; vii) that Federal Government Capital Expenditure on agriculture, as a percentage of total annual budget, does not inspire confidence and high expectation for agricultural production in the years ahead, if the trend remains unchecked; viii) that a major factor often ignored in the down-turn of the agricultural sector in Nigeria is the devastation caused by pests and diseases in the hot and humid environ of ours; ix) the paper concludes by offering suggestions for sustainable plant protection in the new millennium for Nigeria, with a call on government to provide the lacking political will and the needed fillip on plant health matters in a manner similar to those recognized for man and his animals.

1   Introduction

Agriculture is man's oldest profession, not prostitution, as it is often claimed light-heartedly. All human endeavours must derive sustenance from agrarian outputs, wars inclusive. Agriculture is the last hope of man for a new beginning and growth of his institutions. It has been and will continue to be a lead factor in the Nigerian economy for poverty alleviation, income generation and provision of employment, affordable food, clothing and shelter for the country's increasing work force.

A note in the 1998 Annual Report of the International Institute for Tropical Agriculture (IITA) states "—the beginning of civilization depended on agriculture—so does its future".[1] A wise saying asserts—*Primus est edere deinde philosophari*—food first before philosophy, wealth before art. All human endeavours must derive sustenance from agrarian outputs, wars inclusive. Indra de Soysa of the International Peace Research Institute, Oslo, Norway, spoke of a strong parallel between conditions affecting agriculture and poverty, as well as the new forms of global conflicts. Ismail Serageldin, World Bank Vice President for special Programmes, puts it succinctly when he said that it is more cost-effective to avail poor nations technical agricultural assistance now, rather than rushing in refugee foods to war-harassed spots of the globe. Agricultural is the last hope of man for a new beginning and growth of his institutions. It has been and will continue to be a lead factor in Nigeria for poverty alleviation, income generation, and the provision of employment, affordable food, clothing and shelter for the country's increasing work force.

Global Agricultural Production

Agricultural production on a global scale is in a precarious state. A 1977 National Academy of Science (USA) report says that most arable land area of the globe has been fully tilled or farmed. Some 67 % of all pastures are so heavily grazed that they could revert to steppes or deserts within years. Indeed, the size of man-made deserts (1250 million hectares) arising from over-grazing of pastures, indiscriminate exploitation of forest resources and other abuses, is almost equal to that of all tilled land area. And yields are declining in as much as 65% of all irrigated lands due to salt toxicity factors. Grain stocks have been depleting and food prices have risen sharply since the 1970s. Forest resources are being harvested faster than we are regenerating them. Indeed, man has slipped into what has now been aptly referred to as "the Aged of plants" in which, perhaps for the first time in history, almost all almost all known useful plants are in limited supply. On a global scale, however, 25-45% annual losses from the combined effects of pests and diseases on crops have been estimated.[2] For citrus and maize in Africa, losses due to weeds, parasitic plants, disease-inciting microorganisms and unfavourable weather and soil are put at 24 and 69%, respectively. In Nigeria, the hot and humid tropical climate to the south and moistened savannah / fadama enclaves to the north favour a mosaic of pests and diseases. Therefore, by a conservative estimate, we are losing some 50% of our agro-products to pests and diseases, a situation exacerbated by the nation's poor post-harvest management and storage facilities.

The consequences of Nigeria's declining agricultural fortunes have been glaring and far-reaching. The Kano groundnut pyramids have disappeared, cocoa and rubber plantations are being abandoned, while Nigeria's leadership position in palm produce production has been taken over by Malaysia and other nations. The yields of our crops are among the poorest, causing demands for food to be more than supply, with obvious outcome. There is wide-spread famine in the country. It is claimed also that the largest number of malnourished children in Africa, more than 3.5 million, are found in Nigeria. The question has been asked, therefore, why do we as a people suffer excessively from inadequate caloric intake in spite of the apparent potential of our soils and climate to sustain a virile and productive agriculture?

Barriers Facing Nigerian Agriculture

African Agriculture, especially Sub-Sahara Africa (SSA) of which Nigeria is part, faces a series of chronic, acute and interacting barriers. A 1981 report of the International Institute of Tropical Agricultural (IITA)[3] on the subject has summarized these convincingly. They include policy and institutional barriers, limitations imposed by inadequate infrastructure and technology, as well as those arising from the low social standing accorded agriculture in Africa. The IITA report also states that faltering farming systems are among technical barriers of concern that "—offer the greatest challenge to agricultural technology in the humid tropics of Africa." Other constraints facing the Nigerian farmer are storage, quality control and marketing problems, ineffective farm tools and land reform laws, fragile communication and transportation systems, deficiency in research data essential for planning, inadequately funded agro-based research, uneconomical farm size (an average of 0.05 - 4.0 ha in the humid south), marginal input of fertilizers and other farm pesticides, accelerated urban growth and encroachment on agricultural land arising from defective regional planning and distortions in government policies, desertification, floods and horrifying gully erosions. Global Warming has exacerbated these further.

These handicaps are typical for any developing agriculture and of concern in other continents. In SSA, however, there are convincing data to show that the barriers are more accentuated and invidious than elsewhere. It is claimed that the SSA soils are more prone to erosion and degradation than elsewhere. The poor-textured soils with low organic mater content and high acidity respond poorly to inorganic nitrogenous fertilizers, while fertility is diminishing in several parts with serious outcome for sustainable farming.

Absence of production-sustaining farming systems, and a mosaic of uncontrolled pests and diseases induced by a favourably hot and humid weather have exacerbated the situation in SSA to a crisis proportion. Irrigation schemes are sub-marginal in SSA. These factors compel the African farmer to face the whims of unpredictable and excessive rainfalls and the caprices of freakish weather patterns. And as they say, the African farmer is at the mercy of the elements. Institutional and physical infrastructures—including utilities—are epileptic, trained farm hands are in short supply, while urban drift and flight from the farms are said to be more pressing here than elsewhere.

Political opinion varies as to the causes of Nigeria's agrarian backwardness. Some Federal agents in the Alhaji Shehu Shagari Administration in the early 1980's were found of spotlighting what they called the confrontational and non cooperative attitude of some state governments as the major obstacles against achieving self sufficiency in food production. On the other hand, state functionaries, especially from the erstwhile Bendel State, often countered by attributing the cause of food shortages, and hence of escalating inflation in Nigeria, to what they termed as "green manipulation" and by which they meant the shrewd juggling of Federal Government's purse by partisan politicians who exhaust the national treasury through the award of inflated and ill-executed contracts. Time seems to have borne out the fact that in Nigeria there has not always been transparency and accountability in government circles.

It is clear from an appraisal of these barriers cited and many more unmentioned that the prospects for sustainable food production in the next millennium for Nigeria face some formidable challenges. Professor Akin Sodeye, in a keynote address at the 17th Annual conference of the Science Association of Nigeria in 1976, posed a rhetorical question—"How does one set out to change a superstition-ridden cultural environment into a technologically sophisticated milieu"?. [4] The answer, he admitted, is complex, but that it embodies a directional or purposive plan by government and research scientists to influence thinking in that direction. Therefore, let us examine the balance sheet of agricultural production in Nigeria in the past years, as well as the endeavours of government and researchers to date in the regard, an exercise that will give us a clearer view for action in the 21st century.

Since the 1970s government has initiated a series of high-sounding agricultural programmes. The list is impressive, including establishment of state-owned Agricultural Development Programmes (ADP) for extension, direct production parastatals such as the River Basin Development Authorities (RBDAS) and the Directorate of Food, Roads and Rural Infrastructure (DFRRI), Agricultural production programmes such as Operation Feed the Nation (OFN), the Green Revolution, National Accelerated Food Production Programme (NAFPP), Accelerated Wheat Production Programme (AWPP), among other food production companies and a plethora of agro-based Research Institutes, Universities of Agriculture, Faculties and Colleges of Agriculture and Science where researches on agriculture are conducted.

Many of us are members of these organisations, and we are aware of how empty and ineffective they have become. The Universities and Research Institutes have become centres for laughter than for learning and research, for lack of adequate funding, absence of modern working tools and appropriate incentives. The data that follow in relation to food shortages, diminishing yields, farm size and subventions, cannot be explained otherwise, if our numerous research and delivery organs were adequately funded and functioning properly.

Statistics provided by the Raw Materials Research and Development Council (RMRDC) in 1995 on the average national demand and supply for the main raw materials in Nigeria reveal a glaring deficiency in the agricultural sector. All, except only two (soybean and oilseeds—excluding palm) of 21 raw materials are in short supply, with demand outstripping supply by as much a 4×10^6 (wheat), 15×10^6 (rice), 18×10^6 (palm oil) and 80×10^6 (livestock) metric tons. These shortfalls reveal a chasm yearning to be filled by production efforts in our farms. They underscore the instability and non-sustainability in the production cycle, which have occasioned volatile price jumps for agricultural products and below capacity utilization for many agro-industrial enterprises.

Data from the Federal Office of Statistic's (FOS) and the Central Bank of Nigeria (CBN) on estimated hecterages and yields of some major crops in Nigeria, which show a downward trend, give further cause for concern. Between 1980 and 1990, hecterages were down by 15 % (cassava), 29 % (cocoyam) and 69 % for yams. In the same period, yields of Guinea corn, soybean and benniseed went down by 8, 50 and 62 %, respectively. Besides, yields of our major crops are substantially below world averages and those of main producers. A few examples will make the point clearer. Again, for tuber crops, which include cassava, coco yams, Irish and sweet potatoes and yams (*Dioscorea* spp.), the average yield of 9,931kg/ha in Nigeria contrasts discouragingly with 21,454 and 26,783 kg /ha in Japan and the USA, respectively. For rice and maize, our yield levels are 34 and 73 % less than world averages, while yields of these crops in the USA and Japan are 67 and 72 % higher, respectively. A 203 kg /ha average yield for pulses in Nigeria, including broad beans, cowpeas, pigeon peas and peas, contrasts incredibly with 14,471 kg /ha in the USA.

Similar discouraging statistics for diminishing hecterages and sub marginal yields are available for other crops. They portray the poor state of the act in the Nigerian agro-production sector, and signal an urgent need for active research and massive investment in the sector. But, what has been the magnitude of investment in this segment of the economy in the past years.

Federal Government Capital Expenditure on agriculture, as a percentage of total annual budget, does not inspire confidence and high expectations for agricultural production in the years ahead, if the trend remains unchecked. Figures obtainable from the Central Bank show a fluctuation in allocations to the sector from 2.5 % in 1970 to as low as 0.7 % in 1980 and 2.4 % in 1992.

Base data on average annual growth in agriculture and food production in Nigeria during the General Olusegun Obasanjo years (2001-2006) show that the country did not fare any better than some of its West African neighbours. Thus, in the said period food supply and agricultural products increased only marginally by 1.9 and 1.8 %, respectively, compared with 9.3 and 8.5 % growth in war-torned Sierra Loan. Also, during this so-called agricultural golden era for Nigeria when there was so much political chest-thumping for agrarian matters, our import dependency figure was 89 %, compared 94.3 for Sierra Leone. Similar data are in Table 1.[5]

Thus, despite the rhetoric by government and the establishment of an array of agricultural programmes by successive administrations, agriculture remains relegated in budgetary provisions in Nigeria. Governments expressed concern, therefore, has not matched its intentions, as portrayed by its investment profile in the sector.

A major factor often ignored in the down-turn of the agricultural sector in Nigeria is the devastations caused by pests and diseases. Professor M.O. Adeniji, a past president of the Nigerian Society for Plant Protection (NSPP), has highlighted Nigeria's bitter memories of crop failures caused by sporadic outbreaks of pests and diseases in the 1950s - 70s.[1] These include a severe incidence of rust (*Puccinia polysori*) on maize in the 50s, the spread of the noxious Siam weed (*Eupatorium odoratum*)—known commonly now as Awolowo weed in the 1960s, an intense epiphytotic of cassava bacteria blight (*Xanthomonas manihotis*) in the 1960s and 70s, and a caustic defoliant and dieback of yams, but especially water yam (*Dioscorea alata*), incited by the fungus *Glomerella cingulata* that is still very much with us.

Severe threats from other maladies are also still looming. The African cassava mosaic disease (ACMD) is the most devastating disease of cassava is SSA, while a 1982 Progress Report on the IITA's long-range plan says that the cassava mealybug (*Phenacoccus manihotis*) can be considered as African's worst single agricultural pest at the resent time. Together with spider mite *(Monoychiellus tanajoa)*, the pest may cause change in tuber quality, drastic yield reductions and ruinous crop failure for cassava of which Nigeria is the world's largest producer.[5] Other pests in Nigeria are the notorious witch weed (*Striga hermontheca*) wide-spread in sorghum fields, weaver birds (*Quellea spp.*) which are peculiar problems nation-wide in rice and other grain producing areas, but more especially in the North-Eastern parts of Nigeria, and outbreaks of army worms and swarms of grasshoppers (*Zonocerus sp.*) which are frequent and wide-spread. Mirid or leaf miner *(Distantiella theobroma)* pests on cocoa and leaf miner insects *(Coelaenomenodera elaeidia)* on oil palm are with us.[6]

Traditional strategies for the management or control of pests and diseases involve principally the use of biological agents, chemicals, quarantines and resistance breeding. The promise of fungi as pest control agents and the use of other parasites for the biological control of plant maladies are novel strategies for pest control that are gaining acceptance. The symbiotic mycorrhizal fungi found commonly as endophytes in roots of nearly all cultivated crops are pre-eminent in this regard. Two non-mycorrhizal endophytes, *Fusarium oxysporum* and F. *Solani,* have been shown to immobilize banana nematodes and weevils, while *Trichioderma veridae,* an antagonist of *Fusarium spp,* may be manipulated to reduce rot and borer damage in maize. Also, Baevaria bassiana, a mosquito and fly-pathogenic fungus, has provided some promise as a toxin-producing endophyte for stem borer control in maize. The limitations of biological control are the prospects for introducing "Frankenstein monsters" and the fact that many of the promising candidates are yet to be time-tested.

The frequent linkage of resistant genes to those of undesirable traits, often discovered after protracted and expensive breeding programmes, hybrid sterility and pathogen variability that occasion breakdown of resistance, are some limitations in the use of resistant lines to control tf plant depredators. Plant quarantines or exclusion schemes entail the use of plant disease legislations, inspection and elimination of pathogens before distribution or planting. The measures are aimed primarily at non-resident pathogens, such as the restriction of imported propagative cassava materials from the Americas to West Africa to prevent such blastings as "rust" caused by *Uromyces sp.* and "*Oidium*" incited by *Mycosphaerela sp.* that are not yet found in West Africa. The main limitation for quarantine measures is that each country imposes regulations to exclude pathogens of its concern, and there is not always a sound biological ground for the measures adopted, which in turn provide obstacles to international trade.

The control of plants diseases and pest by chemical agents had a historical beginning with the use of copper sulphate by Prevost in 1807 to manage a fungal disease, bunt of wheat, and the formulation of Bordeaux mixture by Millardet in 1885. Since then, some 120 years now, there have been only two major landmark developments in chemical disease control. These are the introduction in 1934 and wide acceptance of the more specific organic or organo-metalic fungicides, pre-eminently the dithiocarbomates, and the wide usage of systemic fungicides since the late 1960s.

The usage of chemical pesticides has become very necessary for sustaining production in modern agriculture. Many crops, such as some vegetables, fruits, tobacco, potato, groundnuts, cocoa, coffee, cereals and cotton, among others, could hardly be produced todaywithout the use of agro-chemicals. Business and economic reports predict a continuing growth or increase in global usage of pesticides, more so in Third World countries which consume only 10% of current world production. It is asserted that catastrophic and famine-generating epiphytotics have been generally averted in the past 100 years or so due mainly to increasing pesticide usage. A 1980 report of the committee on chemical control (CCC) of the International Society for Plant Pathology says that chemical control is, and will continue to be, when other methods fail.

Examples of plant diseases controllable by chemical pesticides are readily cited. The protective chemicals include sulphur that controls effectively the dreaded rubber mildew (*Oidium havaea*) and powdery mildew of grapes *(Uncinular necator)*, dithiocarbamates for potato late blight *(Phytophthora infestants)*, and captan, dodine and dithianon for controlling apple scab caused by *Venturia inaeqalis.* Brown rot of stone fruits *(Monilia sp.)*, loose smuts of cereals *(Ustilago spp.)*, and powdery mildew of apples *(Podosphhaera leucotricha)* and of cereals *(Erysiphae graminis)* are some examples of plant abnormalities manageable by a combination of protectant and systemic fungicides, while decays of bananas caused by *Colletotrichum musae,* of apples by Gloeosporium spp. and of oranges by *Phomopsis citri* and *Diplodia natalensi* are among plant abnormalities that have been effectively controlled by systemic fungicides, such as benomyl, imazalil and thiobendazo.

These inroads made in chemical control towards sustainability in agriculture are far from saying that all is satisfactory with pesticide usage schemes. There remain a number of sensitive issues to be resolved. First, there are some diseases that cannot be controlled adequately by chemical pesticides from available information today. These include *Verticillium* wilt of some perennial crops and witches broom disease of cocoa caused by *Crinipellis perniciosos*. Some others, such as those caused by *Armillaria, Phytophthora* and nematodes, are controllable only by heavy soil fumigation or undesirable repeated pre-planting applications. Additionally, some bacterial, viral and even fungal pathogens, such as *Alternaria* and *Plasmodiophora*, are not affected by many of the known chemicals. Another limitation involving the use of chemicals in agriculture is the phytotoxic side effects they cause. Ethylenebis dithiocarbamate or zineb is injurious to some pear varieties, while Bordeaux mixture or copper sulphate formulations are less popular now than organic fungicides as candidate pesticides for managing late blight on tomatoes and potatoes due to their damaging effects on these and other crops. Other damaging manifestations of chemical pesticides are reduction in yields, storage or market quality for some crops in certain situations.

A major disincentive against the use of agricultural chemicals is the hazards they are known to cause on non-target organisms (humans inclusive) and on the environment—two highly emotional or controversial issues. An extreme instance of a fatal poisoning by a pesticide was the Bhopal disaster in India in late 1984 in which some 1200 humans and thousands of domestic and wild animals died from an underground gas leakage from an insecticide plant belonging to the Multinational Company, Union Carbide. Another catastrophic incidence of poisoning from a chemical pesticide occurred in Iraq in 1972 from the ingestion of bread laced accidentally with a methyl mercury fungicide supplied to farmers for seed treatment. Some 6,350 persons were hospitalized, with 459 deaths.

Nevertheless, it is to be noted that instances of pesticide poisoning are rare, mainly accidental or due to suicide or crime. In fact, the CCC concluded that the danger of accidental poisoning from most disease control chemicals is negligible and that "normal intake of pesticides in foods is unlikely to present any hazard", due to the conservative safety margins often used in delineating optimum safety limits for pesticides. Furthermore, the committee stated that, although the issues of environmental pollution occasioning pesticide residues in the air, soil and water are even more contentious now, there is no evidence that most disease-control chemicals cause harmful environmental alterations.

Perhaps the most serious limitation affecting the economic usefulness of farm chemicals is their known induction of pathogen resistance. This is a situation in which resistant races of pathogens previously unknown may arise after pesticide usage with time, a costly and disturbing phenomenon for farmers and pesticide producers. Resistant races of pathogens have been discovered against several pesticides, such as *Alternaria kikuchiana against polymaxin, Pyricularia oryzae (kasugamycin),*

Erwinia amylovora (streptomycin) and of several other pathogens against the benzimidazoles. The use of the novel antipathogenic agents—effective chemical theraputants that are seemingly non-toxic to pathogens—is suggested as one way out of this problem. It may however be concluded that pathogen resistance to agro-chemicals will be a main setback to pest and disease control efforts in the 21st century, engendering non-sustainability.

The pesticide era of the 1930s and 1940s, during which period such miracle pesticides as DDT (insecticide), 2-4-D (herbicide); Bordeaux mixture (fungicide), penicillin (bacteriocide) and several others were introduced and used widely, derived mainly from the threats of the biotic disease-causal agents on our crops and animals. However, the successes achieved by using chemical pesticides have not been as great as the zeal with which they were introduced.

Nevertheless, chemical pesticides are still the most important weapons in the arsenal of plant protectionists. And when used in combination with other measures, such as resistance breeding, biological cultural control methods, intelligent use of chemical pesticides is an ecologically sound and an essential part of the modern integrated pest management (IPM) schemes.

Perhaps the main setback in the development and implementation of IPM programmes is a lack of the political will on the part of the Nigerian government to provide the needed uplift on plant health matters in a manner similar to those recognized for man and his animals. The preciousness of the human life and the monetary value of domestic animals have always compelled governments to provide massively for human and animal health care programmes, while plants always played the third fiddle.

Suggestions for sustainable plant protection in the new millennium

The high population growth rate in Nigeria, now put at about 3% per annum, needs a proper management. A shift in emphasis from energy and cost-intensive inorganic nitrogen fertilizers to organic and biological nitrogen fixation may reproach our soil fertility problems. Reforestation and other ecologically sound reclamation programmes can halt the desertification, erosion, flooding and pollution of the limited land resources. Some level of mechanisation of farm operations is inevitable in sustainable farming in the 21st century for Nigeria. A consolidation of our political independence, emphasis on agriculture and economic matters, and a re-ordering of priorities with a fillip on agricultural research, education and energy are the answers. Research enables us to define problems and provide strategies for controls. A clear perception of the importance of plant health problems and a clear definition of the extent and magnitude of crop losses are needed through research to convince administrators and policy makers of the need for more resources to combat plant handicaps.

Actual and potential dangers of mycotoxins and hazards in foods need proper evaluation through research to emphasize quality-reducing effects of molds and other organisms. Manpower requirement for plant health schemes need a survey, as well as a commitment on the part of government to meet these requirements. Appropriate incentives for plant protection crews are needed to motivate these categories of works so as to avoid brain drain presently estimated at more than 20 per cent in Nigeria. Progressive legislations should be made to moderate uncontrolled usage of environmentally-hazardous pesticides, halt their indiscriminate sale by unlicensed agents, and end the era of making Nigeria a dumping ground for pesticides outlawed in other countries. As integrated pest management

(IMP) is the control strategy in vogue, collaborating crop improvement teams—including plant pathologist, insect scientists and plant breeders, among others—are essential for effective evaluation of germplasm collection of plants for resistance against an array of hazards. Agronomists are wanted to march crop cultivars with seasons and localities. Properly trained extension specialists—*Doctors of Plant Medicine—currently* estimated at one for every 1000 farmers in Nigeria—are invaluable for the dissemination of useful and practical information on plant health problems, and for sound and expeditious implementation of research results. These suggestions underscore a need for comprehensive training programmes, a national plant health schemes and a world health organization for plants, similar to those for humans and animals.

Table 1. Average annual growth in agriculture and food production in Nigeria and other West African countries in 2001-2006 (% of 1999-2000 figures).

Agricultural products:	1.8 % (Nigeria), 2.5 % (Ghana), 8.8% (Sierra Leone).
Food supply:	1.9% (Nigeria), 2.6% (Ghana), 9.3 % (Sierra Leone).
Cereal production:	4.2, 3.7 and 3.6%, respectively.
Self sufficiency ratio:	89.7, 84.2 and 94.3 %, respectively.
Import dependency ratio:	10.4, 16.6, and 5.8 %, respectively

Source: Statistics Department, African Development Bank, Tunis, Tunisia (2010).[7]

REFERENCES

1. International Institute of Tropical Agriculture (IITA). 1998. Annual Report for 1998. IITA, Ibadan, Nigeria. 94 pp.

2. Adeniji, M. O. 1977. Status of plant protection in the Operation Feed the Nation Programme, p. 14-20. In *Occasional Publication No. 2*, Nigerian Society for Plant Protection, University of Ife, Moor Plantation, Ibadan, Nigeria.

3. International Institute of Tropical Agriculture (IITA). 1981. Annual Report for 1981, IITA, Ibadan, Nigeria. 178 pp.

4. Sodeye, A. I. 1976. The journey from Nok Culture to space age technology, p. 180-183. In A. Youdeowei and O. Dipeolu (eds.), Bulletin of the Science Association of Nigeria, vol. 2. Science Association of Nigeria, Ahmadu Bello University, Zaria.

5. Ayanru, D. K. G. 1987. Effects of mealy bug (*Phenacoccus manihoti)* infestation on cassava yield components and plant tissue quality. *Der Tropenlandwirt* **88**: 5-10.

6. Hartley, C. W. S. 1979. Integrated plant protection in oil palm, p. 549-552. In Thor Kommedahl (ed.), *Integrated Plant Protection for Agricultural Crops and Forest Trees.* Burgess Publishing Company Minneapolis, Minnesota.

7. African Development Bank (ADB), Statistics Department. 2008, Selected Statistics on Africa. African Development Bank, Tunis, Tunisia. 300 PP.

7.10 Poverty Burden, Energy Consumption And Environmental Stress: A Peep Into The Use Of Biomass Fuels In Sub-Saharan Africa.

G.E.D. Omuta[1]

1 Introduction

Maintaining a sustainable balance between basic human needs and the exploitation of natural resources, on the one hand, and the regenerative capacities of the environment, on the other, could be a very huge challenge. Dealing with environmental matters requires a good understanding of the relationships between and among man, his physical habitat, his socio-economic status and the processes that produce peculiar environmental challenges; since given populations tend to make peculiar demands, and create peculiar impacts, on given environments. In other words, environmental challenges are the results of the interplay of physical, social, economic and political factors of the human processes of production, consumption and social reproduction (Lowe and Bowlby, 1992).

Since the late 1960s, African leaders have recognized the importance of the environment and its resources to the development of their economies. Consequently, they have taken steps to protect and conserve their environments. Some of these steps include the adoption of:

- ♣ The African Convention on the Conservation of Nature and Natural Resources; the so-called *Algiers Convention*, in 1968 (OAU, 1968);
- ♣ The 1991 *Bamako Convention* on the Ban of the Import into African and the Control of Transboundary Movement and Management of Dangerous Wastes within Africa (Amechi, 2009);
- ♣ The *Nairobi Convention* for the Protection, Management and Development of Marine and Coastal Environment of the East African Region (www.unep.org/Nairobi Convention/docs/ English_Nairobi_Convention_Text.pdf); and
- ♣ The Convention for Cooperation in the Protection and Development of the Marine and Coastal Environment of the West and Central African Region (http://www.opeworg/chemica-weapons-covention/related-international-agrrements/toxic-chemica-weapons-and-the-environment/marine-and-coastal-environment-west-and-central-africa/).
- ♣ The *Nairobi Convention* for the Protection, Management and Development of Marine and Coastal Environment of the East African Region (www.unep.org/Nairobi Convention/docs/ English_Nairobi_Convention_Text.pdf); and
- ♣ The Convention for Cooperation in the Protection and Development of the Marine and Coastal Environment of the West and Central African Region (http://www.opeworg/chemica-weapons-covention/related-international-agrrements/toxic-chemica-weapons-and-the-environment/marine-and-coastal-environment-west-and-central-africa/).

---

[1] Gideon E.D. OMUTA, Professor Geography and Regional Planning, Department of Geography and Regional Planning, University of Benin, Benin City, NIGERIA

At the individual national levels, a myriad of environmental instruments have also been adopted to protect the environment so that it can stimulate, enhance and continue to sustain socio-economic development. These include:

- ♠ South Africa's National Environmental Management Act (NEMA), No. 107 of 1998,
- ♠ Kenya's Environmental Management and Coordination Act (EMCA), No. 8 of 1999, and
- ♠ Nigeria's Federal Environmental Protection Agency (FEPA) Act, Cap F10 *LFN* 2004, now superseded by National Environmental Standards and Regulations Enforcement Agency (Establishment) Act, No. 25 of 2007.

Apart from these continental and national environmental instruments, Amechi (2009) has also noted various environmentally-inclined human rights instruments aimed at the promotion of socio-economic development in Africa. He specifically cited article 24 of the African Charter on Human and People's Rights; known as the ***Banjul Charter***, which provides that "all peoples shall have the right to a generally satisfactory environment, favourable to their development" (OAU, 1982, in Amechi, 2009: 109). Twenty-seven (27) African countries have followed the Banjul Charter by explicitly enshrining environmental protection and conservation in their constitutions. A sample will suffice.

- ❖ The 1992 constitution of Angola states that "all citizens shall have the right to live in a healthy and unpolluted environment". Part II, Article 24(1) directs the state to "take the requisite measures to protect the environment and natural species of flora and fauna throughout the national territory and maintain ecological balance" (Earth Justice Report, 2005: 86),
- ❖ The 1990 constitution of the Benin Republic states that "everyone person has the right to a healthy, satisfying and lasting environment" (Earth Justice Report, 2005: 88),
- ❖ The 1992 constitution of the Republic of Congo states that each citizen shall have right to a healthy, satisfying and enduring environment. Title II, Article 46 directs the state to "strive for the protection and conservation of the environment" (Earth Justice Report, 2005: 91),
- ❖ The 1996 constitution of the Republic of South Africa states that "everyone has the right to an environment that is not harmful to their health or wellbeing", while Chapter 2, Article 24 directs the state "to have the environment protected for the benefit of present and future generations, through reasonable and other measures that secure ecologically sustainable development and use of natural recourses while promoting justifiable economic and social development" (Earth Justice Report, 2005: 104).

This review shows that the stubborn persistence of environmental challenges cannot be wholly attributed to the lack of regulatory policy instruments. Consequently, Amechi has asked what then is (are) the cause (causes) of environmental stress and degradation in Africa, and how can they be tackled so as to realize the various environmental goals and objectives as expressed in the various extant continental, national and human rights environmental instruments?

In order to attempt to answer these questions and achieve the objective of this contribution, the rest of the paper is divided into four sections. First, the concept of poverty is outlined in the context of the paper. Then, the manifestations of poverty in meeting the energy needs of, particularly, rural Africa are presented. The links between energy needs and the environment are briefly discussed. The challenges that these manifestations present to policymakers are discussed. The policy implications of the paper are outlined. Finally, concluding remarks are made.

2   POVERTY AS A COMPLEX PHENOMENON: A CONCEPTUAL DISCOURSE

The debate about poverty-environment nexus dates back to the 18th century, when Thomas Malthus argued that because of their inherent economic handicap, the poor "seldom think of the future" and consequently continually degrade their natural recourse base and put their environment under perpetual pressure and stress (Malthus, 1798). In contemporary context, Ostrom, Burger, Field, Norgerd, and Policansky (1999) argue that people in poverty are forced to deplete resources to survive, and this degradation of their environment further impoverishes them. Similarly, Holden, Bekele and Pender (2004) and Agrrey, Wambugu, Karugia and Wanga (2010; 84) have argued that poverty-constrained options may, and do often, induce the poor to deplete environmental resources at rates that are incompatible with the long-term sustainability.

However, in order to be able to properly understand the connections and deal with the issues pertaining to them, the concepts of poverty and environment must be investigated. First, what is poverty? Is it a single attribute? Or is it multifaceted? If it is multifaceted, what are its dimensions and components? In this section, we shall explore and characterize the concept of poverty.

Economists have traditionally defined poverty on the basis of household incomes or consumption capacity, and have adopted this as the best proxy for welfare (Bucknall, Kraus and Pillai, 2000). This perspective confines the definition of poverty to 'those areas of life where consumption and or participation are determined primarily by command over financial resources' (Nolan and Whelan, 1996: 193). Consequently, using this approach, people are considered poor if their levels of consumption fall below a given income (poverty) line, which is currently globally set at one US dollar per person per day. However, Langmore (2009) has been cited as arguing that such a narrow perspective to the definition of poverty implicitly excludes non-material elements found in broad United Nations definitions, such as 'lack of participation in decision making', 'violation of human dignity', 'powerlessness', 'susceptibility to violence' and 'humiliation' (www.policy.co.uk/key concepts/samples/lister-chapter.pdf). In other words it is now clear that poverty is a multi-dimensional phenomenon.

It is from this point of view that Chukwu (2008) has defined poverty as "a condition of lacking the necessary ingredients that make life worth living", to the extent that these ingredients could be many and quite varied; material and non-material. In other words, it is difficult to find a single measure for the analysis of poverty. Therefore, contemporary definitions of poverty are moving from single dimensions to include utility- and capability-based concepts. These include inequality (both within a country or region and within a household), health, education, security, energy use, political voice, and discrimination (Sen, 1981; Putnam, Leonardi and Nanetti, 1993). It must be noted that not all of these indicators are necessarily applicable in every particular situation, but generally, each is relevant in capturing an aspect of the global, composite concept of poverty (Ravallion, 1996).

In the view of Nolan and Whelan (1996), the definition of poverty lies in whether its conceptualization is rooted in the individual's personal resources, especially income, on the one hand, or in terms of living standards and activities; while Ringen (1987: 146) believes that both could be combined, and has consequently defined poverty as "a low standard of living, meaning deprivation in way of life because of insufficient resources to avoid such deprivation".

The definition of poverty has also been dichotomized into 'absolute' and 'relative'. In this regard, absolute poverty has been defined in terms of an individual's lacking sufficient resources to meet his or her basic physical needs. It is translated ultimately into survival or more commonly, subsistence; linked to basic standards of the physical capacity necessary for production (paid work) and reproduction (the bearing and nurturing of children) (Joseph and Sumption, 1979: 27). According to the United Nations Fund for Population Activities (UNFPA) absolute poverty is "a condition of life so limited by malnutrition, illiteracy, disease, squalid surrounding, high infant and low life expectancy as to be beneath any reasonable definition of human decency" (1991; 65).

Relative poverty, on the other hand, derives from Townsend's monumental work: "***Poverty in the United Kingdom***", and puts poverty within its social context and consequently declares that:

> "Individuals, families and groups in the population can be said to be in poverty when they lack the resources to obtain the type of diet, participate in the activities and have the living conditions and amenities which are customary, or are at least widely encouraged and approved, in the societies to which they belong. Their resources are seriously below those commanded by the average individual or family that they are, in effect, excluded from ordinary living patterns and activities" (Townsend, 1979: 31)

The point being made is that the discourse of the poverty burden has become very robust and more dynamic and has been broadened to include investigations of the parameters of its various indicators.

3   Energy Consumption By The Poor

One of the indicators that could be, and has been, used to measure poverty in developing countries is the household's affordability of environmentally-friendly sources or types of energy (or what has been called 'clean' or 'modern' energy) for domestic use. In this context, clean or modern energy may be defined as any energy that either does not pollute at all, or the one that, while reducing the amount of pollutants put into the world, also comes from a source that does not compromise the ability of future generations to meet their energy needs (or what has been called 'environmental debt') (www. tinygreenbubble.com/livegreen/.../546-clean-energy-definition;   dictionary.reference.com/browse/ clean + energy; www.wisegeek.com/what-is-clean-energy.htm; ODI, 2006). Clean energy is needed to save the world from climate change and ensure future energy security (file:///C:Users/Owners/ Desktop/clean_energy_definition.html). Common forms of clean energy include solar energy, wind energy, hydro energy, even rain energy (www.Twitter.com/TinyGreenBubble; www.wisegeek.com/ what-is-clean-energy.htm).

One thing that is common to all forms of clean energy is that the technologies needed in their production processes and commercialization are alien to most developing countries because they tend to be very sophisticated and or complex. Consequently and expectedly, they are either not commonly available, or where they are available, are expensive and unaffordable to the poor. There is, therefore, a strong relationship between income, which is a common parameter for measuring poverty, on the one hand, and the type of energy consumed, on the other. In this connection, for instance, studies by Gundimeda and Kohlin (2003) have shown that while wood fuel is accepted as a "normal good" for the poor, it is considered an "inferior good" for the high income households. In the context of Nigeria, Adeoti, Idowu and Falegan (2003) have found that the relationship is such that the use of fuel wood contributes to poverty because the vulnerable, poor households spend almost twice as much on fuel wood for cooking as the mean national income. In Lesotho, Wason and Hall (2004) and Scott (2006) found that when the government increased the price of oil-based (clean, modern) fuels, many Basotho decided to economize, by returning to using traditional (dirty) fuels, despite the fact that fuel wood was already becoming very difficult to find.

To further appreciate the poverty and energy connection, we must note the accepted fact that population growth is faster among the poor. It must further be reiterated that the poor have tended to be almost always dependent on their environmental resources base for their basic needs, including energy. As Osei-Hwedie (1995) puts it, as population expands and the number of the poor increases, demands on resources will also increase; for example, the demand for food, cooking fuel and wood would put greater pressure on agricultural land as well as the stock of a number of environmental resources, especially traditional sources of energy.

Biomass is the composite name for energy from such sources as wood energy or wood fuel or fire wood, animal dung, agricultural wastes and residues, and charcoal. They constitute the bulk of the energy sources for the poor and they are the ones usually classified as 'dirty' (file:///C:Users/Owners/Desktop/clean_energy_definition.html).

It must be noted from the onset that the collection and processing of biomass energy statistics are very complex. This is because of the diversity of consumption patterns, differences in the units of measurement, the lack of regular surveys and the variation in the heat content of different types of biomass, among others.

In terms of the global scenario, and according to the best available data (WEO, 2006; www.worldenergyoutlook.org/docs/weo2006/cooking.pdf), household energy use in developing countries totaled 1,090 Mtoe (million tons of oil equivalent) in 2004, which is almost 10 per cent of the world primary energy demand. While in the Organization for Economic Cooperation and Development (OECD) countries, the bulk of the energy demand comes from the power generation and industry sector, in the developing countries, these sectors represent only 12 per cent.

According to World Energy Outlook (WEO, 2006; 420),although there are enormous variations in the levels of consumption and the types of fuels used, and although a precise breakdown is usually difficult, the main use of energy in households in developing countries, including, sub-Saharan Africa is for cooking, followed by heating and lighting. According to the Aprovecho Institute of Germany, the average poor family in developing countries uses about four tons of fuel wood annually for cooking, which is about twice as much wood as each family in the developed world uses for construction, paper, furniture and fire wood, combined. The Institute concludes that up to 70 per cent of all the wood used on earth ends up under someone's cooking pot (Aprovecho Institute, 1984; 5). Consequently, in this contribution, our emphasis is on cooking, being the main energy demand of the household in sub-Saharan Africa.

Indeed, the Duetsche Gesellschaft fur Technische Zusammenarbeit (GTZ, 2007) has confirmed that in developing counties, and especially in the rural areas, 2.5 billion people rely on biomass sources of energy, such as fuel wood, charcoal, agricultural waste and animal dung (biomass), to meet their energy needs for cooking. Furthermore, these resources account for over 90 per cent of the aggregate household energy consumption (WEO, 2006; 419; www.gtz.de/de/dokumente/en-cooking energy-2007.pdf; 3; IEA. 2006). Indeed biomass fuels are often the only available energy source, especially in rural areas. And in most sub-Saharan African countries, it has been confirmed that more than 80 per cent of the population use, and will continue to use biomass fuels for their daily cooking (www.gtz.de/de/dokumente/en-cooking energy-2007.pdf; 3). In fact, it has been projected that in the absence of new strong and responsive policies, the number of people that will rely on biomass to meet their energy need for cooking will increase to over 2.6 billion by 2015 and to 2.7 billion by 2030, due largely to population growth and stagnant or sluggishly responding per capita income (WEO, 2006; 431).

Probably more importantly, the United Nations Millennium Project (2005) has established close and strong links between energy needs, on the one hand, and each of the eight (8) Millennium Development Goals (MDGs), on the other. Thus, it is argued that a shift from traditional to modern energy sources helps to reduce poverty (MDG 1); and by freeing the vulnerable population segments, can improve educational opportunities for children, empower women and promote gender equality (MDGs 2 and 3). Furthermore, the availability of clean and safe energy (in contrast to the hazardous traditional sources) is considered important in reducing child mortality (MDG 4). It is also argued that freeing women from the burden of carrying heavy loads of fuel wood reduces maternal mortality (MDG 5). More efficient and complete combustion of fuel wood reduce respiratory illnesses and other diseases (MDG 6).Also it is argued that fuel substitution and improved stove efficiencies would help alleviate the environmental damage of biomass use (MDG 7). Finally, widespread substitution of modern types and sources of energy for the more common traditional biomass will engender and facilitate global partnerships (MDG 8) (WEO 2006; 430-431).

Having established that cooking is the main use of energy in developing countries, the above scenario helps us to understand its strategic place in the development of such regions.

Generally, in developing countries, households use a combination of energy sources for cooking. These can be categorized into:

- ❖ Traditional; including: dung, agricultural wastes and residues and fuel wood (wood energy),
- ❖ Intermediate, such as charcoal, and
- ❖ Modern; including: liquefied petroleum gas (LPG), biogas and electricity.

In developing countries, including sub-Saharan Africa, electricity, biogas and LPG, together, represent a very insignificant proportion of total household energy consumption, because, being largely either unavailable or unaffordable, they are mainly used for lighting (and, where appropriate, for powering light appliances), rather than for domestic cooking. Bearing in mind also, that our concern is with the poor in these countries, electricity and other modern sources of energy are not our focus. Considering further, the very close affinity between charcoal, on the one hand, and agricultural residues and fuel wood, which make up the bulk of the traditional energy source, on the other, we shall not make a distinction between traditional and intermediate sources of energy.

In stating the importance of traditional sources of energy in many developing countries, Ouedraogo (2005) has argued that the high cost of modern cooking energy such as LPG and electricity and their cooking stoves are major constraints for household fuel preferences. Similarly, Fawehimi and Oyerinde (2002) have used descriptive statistics to confirm that massive increases in the costs of modern sources of energy have resulted in correspondingly increasing levels of relative poverty in Nigeria. Consequently, they concluded that wood energy has remained, and would continue to remain, the fuel of choice for most homes. In his regard, Chavin (1981) in Ouedraogo (2005) and Arnold, Kohlin Presson and Shephard (2003) have confirmed that in spite of several public policies designed by the government of Burkina Faso, to discourage the continued dependence on wood energy, "a certain inertia is observed for household cooking fuel preferences". In his study of Burkina Faso, Ouedraogo (2005) confirmed that more that 90 per cent of the population of the country use wood energy as the main source of cooking energy and that even in urban Ouagadougou, the national capital, only 29 per cent of the population has access to electricity, which is used exclusively for lighting. The study further confirmed that the poorest segments of the society which have the largest families and cook more food and more frequently, are the main users of fire wood.

The studies by Tolba, El-Kholy, Holdgate, McMicheal and Munn (1992) confirmed that in Kenya, 78 per cent of the population depend on fuel wood for cooking, while in Ethiopia, the figure rises to 95 per cent. In Tanzania, the figure is as high as 95 per cent while, in rural Botswana the proportion of households that use fuel wood for cooking is three times that of urban areas (IEA, 2006; 422). With crises and shortages already manifesting, Duraiappah (1996) has expressed serious concern for the energy security of the poor, who depend solely on this source of energy for cooking and warming. Probably of equally, if not more significant concern are the implications of these pressures and the attendant consequent shortages for the environment from which the cooking energy is sourced and in which it is used.

Energy Consumption Of The Poor And The Environment

A plethora of definitions of the concept of the environment exist in the literature, ranging from the sweeping and general, to the sophisticated and specific. Hence, Comim (2008; 6) posits that "controversies exist about the concept of environment" because it covers a wide range of ecological aspects. In colloquial usage, the environment is defined as our surroundings, especially the material and spiritual influences which affect the growth, development and existence of a living being (The Concise Oxford Dictionary, 1995). This is in consonance with Wilkinson and Wyman's (1986) definition of the environment as all the interesting factors and circumstances that surround influence and direct the growth and behavior of individual beings, groups, species and communities. Earlier, Detwyler and Marcus (1972) had defined the environment as the aggregate of the external conditions that influence the life of an individual or a population. A logical extension of these definitions is that since the mix of external conditions almost always varies from place to place, the quality of the environment will correspondingly vary from place to place (Omuta, 1988, 418).

In the opening remarks, we noted that African countries have long been concerned about protecting and conserving their environmental resources, as demonstrated by their declarations in the African Charter on Human and Peoples Rights. For instance, countries like Angola, Benin and South Africa have enshrined in their various constitutions, their commitment to protecting the health of their people (even the flora and fauna) and guaranteeing an unpolluted environment.

In this segment, we shall examine some of the effects of the demand for biomass energy on the quality of the physical surroundings (the macro environment) of the source regions, on the one hand, and how the traditional use of biomass affects the immediate domain (micro environment) of the user (Omuta, 2004; 7), on the other.

Macro Environmental Effects of Biomass Energy Consumption

Inefficient and unsustainable cooking practices can, and do in fact, have serious implications for the physical environment. These impacts could manifest in various forms including land degradation, as well as local and regional air pollution. Tolba, El-Kholy, Holdgate, McMicheal and Munn (1992) have claimed that in 1980, 1.3 billion people were already facing fuel wood shortages because the rate of consumption was no longer sustainable. Furthermore, they raised an alarm that if the (then) rate of deforestation continued, the shortages would escalate and the figure would have reached 2.7 billion by the year 2000. If these projections were near accurate, the present levels of shortages can only be described as catastrophic. Ouedraogo's (2005) study shows that in Burkina Faso, the urban population of Ouagadougou have known wood energy crisis since the 1970s, especially during the rainy season, when fuel wood scarcities are frequently experienced. Sow (1990) noticed that many people today, do not have enough wood to cook more than one meal per day, and for such people, the real energy crisis is the lack of wood energy. With a projected probability of the adoption of fuel wood as the major source of cooking energy put at 92.2 per cent, Ouedraogo (2005) was lead to conclude that "wood fuel will remain the primary source of energy for households in Ouagadougou for a long time to come". The implication is that the longstanding shortages will not only persist but should be expected to escalate.

However, the debate as to how much pressure the balance between wood fuel demand and supply places on forest resources appears to be inconclusive. Indeed, the positions that have been taken can be poles apart. This is so because some are of the opinion that the unsustainable preference for fuel wood portends a great threat to the forest, the users and the economy, and that indeed, the depletion of resources has resulted in an increase of direct and indirect costs (Ouedraogo, 2005; FAO, 2003). In fact, Somaratne (2005) has argued that urban residents' demand for fire wood as energy source leads to rural deforestation.

Contrastingly, others have argued that national intervention is not necessary because it has not been established that wood fuel demand is a major cause of deforestation (Arnold, Kohlin Presson and Shephard (2003). The argument of the latter group is that as much as two-thirds of the fuel wood used for domestic cooking is usually gathered from the road side and trees outside the forests rather than from natural forests (IEA, 2006).

The unresolved nature of the debate notwithstanding, some strong and seemingly emphatically vehement positions have been taken. For instance, GTZ (www.gtz.de/de/dokumente/en-cooking energy-2007.pdf; 5) claims that in the many cases where the demand for biomass fuels far exceed sustainable supply, the inevitable result is "massive deforestation, land degradation and desertification".

Amechi (2009) has tried to establish a more direct relationship between the energy need of the poor and environment. In this regard, he argues that the poor is often forced to rely heavily on the ecosystem for his energy needs, which in most cases leads to its degradation. Similarly, Johnson (1993; 145) argues that the poor are driven to destroy the environment because they have no other possibilities to explore. For them, it is a question of sheer survival. Amechi (2009), therefore, concludes that in sub-Saharan Africa, the issue of deforestation exemplifies the effect of income poverty on the environment, where the rising demand for fuel wood and charcoal for energy needs has been identified as one of the major causes of deforestation in the region. He attributes the rising demand to the inability of the poor to access modern and cleaner energy sources because of their lack of income. IEA (2006) has established that while fuel wood may not be gathered directly from the forest, charcoal, which is also a major component of the cooking energy used by the poor, is usually produced from forest resources. Furthermore, it has been argued that unsustainable production of charcoal in response to urban demand, particularly in sub-Saharan Africa, has placed serious strain on biomass resources. It is also argued that because charcoal production is often inefficient, it has lead to localized deforestation and land degradation around some urban areas. For instance, charcoal production for urban and peri-urban households has resulted in the devastation of biomass resources over a radius of 200 to 300 kilometers around Luanda, the capital of Angola (*ibid*; 427). In their study of Zambia, Liberty and Hongjuan (2008; 373) established that the need for fuel wood "is one of the major causes of the reduced tree cover in Magoye West", where excessive lopping and felling combined with the poor regeneration capabilities of trees and the increasing commercialization of the fuel wood and the charcoal economy have accelerated the rate of deforestation.

Given the low level of technology adoption and patronage in sub-Saharan African countries, one of the ways that small scale and peasant farmers have been regenerating their farm lands is to allow animal dung and farm residue to decompose *in situ*. However, since their incomes either limit or prevent their access to cleaner energy sources, the poor are constrained to use these natural fertilizers as cooking fuel. When animal dung and farm residue are used for fuel rather than left in the farm or ploughed back into the fields, soil fertility is reduced and propensity to soil erosion is increased. The end result is general land degradation.

Within the bigger deforestation picture are some other equally important aspects or sub-sets of environmental degradation due to unsustainable biomass energy demand. These include biodiversity loss as well as atmospheric pollution. Regarding atmospheric pollution, it has been reported that in Kenya, charcoal production and consumption emit more greenhouse gases (GHGs) than the industries and transportation sectors put together (RK, 2002).

Micro Environmental Effects of Biomass Energy Consumption

Our concern here, is with the environment in which biomass is converted into energy and used for cooking; namely, the home. For the poor in sub-Saharan African countries, the home is usually a small space, which could be as small as one poorly ventilated and over-crowded room for the entire family. The housing of the poor is usually also characterized by the absence of kitchen facilities, or where there are kitchens, they are shared and are without cooking gadgets, except the ubiquitous three-stone fire stove. In the most common cases where there are no kitchens, cooking is done indoors; which could be inside the same room that the family sleeps in, or along the corridor that leads to many rooms. In all cases: whether they are along corridors or inside the room, the conversion of biomass into cooking energy by the poor results in indoor pollution. The pollution is as a result of either incomplete combustion of the biomass or the non-dissipation of smoke or both.

The most important negative environmental externality arising from air pollution due to indoor cooking, using biomass is health deterioration (Ezzati and Kammen, 2001). Smith *et al* (2000) have established that biomass produces high emissions of carbon monoxide, hydrocarbons and particulate matter (PM). Furthermore, it has been well established that indoor pollution arising from biomass-based indoor cooking is a primary contributor to respiratory problems (Ezzati and Kammen, 2001). Indeed, UNEP (2006; www.unep.org/pdf/annualreport/UNEP_AR_2006_English.pdf) has estimated that air pollution causes about 36 per cent of lower respiratory infections and 22 per cent of chronic respiratory diseases. The World Bank (1992) has also asserted that between 300,000 and 700,000 deaths can be prevented if they are removed from unsafe levels of indoor smoke, or if suspended particulate matter (SPM) concentrations can be held to WHO standards.

The most vulnerable groups that suffer from indoor air pollution (population at risk) are the women and children, who invariably and traditionally do most of the cooking and other household chores and are consequently exposed for long hours to smoke (DFID, 2000). Suspended particulate matter concentrations are high because biomass fuels are usually converted into cooking energy by burning in inefficient open fires and traditional stoves. In fact it has been claimed that every year, the smoke from open fires and traditional stoves kills 1.5 million people. In other words, every 20 seconds, a woman or child is dying due to inefficient use of biomass fuel (www.gtz.de/de/dokumente/en-cooking energy-2007.pdf; 5). An even more alarming scenario was painted by the World Bank (2001) which explained that smoke inhalation results in severe health problems and that it contributes to about 4 million deaths per year among infants and children in developing countries. Elsewhere, the World Health Organization (WHO) has estimated that more than 4,300 deaths of children below five years of age per day are due to biomass energy use. This means that indoor air pollution associated with biomass energy use is directly responsible for more deaths than malaria; almost as many as tuberculosis and almost half as many as HIV/AIDS (IEA, 2006; 425). UNEP (2006) has established that a child exposed to indoor air pollution is two to three times more likely to catch pneumonia, which is one of the world's leading killers of young children. Furthermore, IEA (2006) asserts that there is evidence to link indoor smoke inhalation to low birth weight, infant mortality, tuberculosis, cataracts and asthma

Among the indirect social and economic negative externalities of the health implications of cooking indoors with biomass fuels is the low productivity among adults and mental retardation among children. In this regard, it has been estimated that urban centres which have SPM levels above the WHO standards lose an equivalent of 0.6 to 2.1 working days per year for every adult in the labour force due to respiratory related illnesses (World Bank, 1992). Although no confirmatory studies have been done, it is estimated that the medical cost burden on the economy could be staggering, with consequential implications, especially for the poor economies of sub-Saharan Africa.

In addition to time-loss due to ill health, it has also been argued that dwindling biomass fuel resources leads to additional workload for the women and children whose lot it is traditionally to source cooking fuels. The distances covered continue to increase, as the energy sources become increasingly scarce, with the implication that they spend correspondingly increasingly more time searching for fuel wood. This means less and less time for other productive, income-earning activities, even including the education of the children. For instance, in Tanzania, a national survey of 22,178 households showed that in the central region of Singida, people travel an average of 10.4 kilometers daily (one direction) to collect fuel wood (IHSN, 2006). The heavy loads of fire wood carried over such long distances, in turn, have implications for the health and productivity of women and children.

4   Matters Arsing

Among the negative externalities associated with the consumption of biomass fuels by the poor are that:

♣ Biomass fuels are converted into cooking energy by burning, using very inefficient and very ineffective methods, such as three-stone open fires and traditional stoves,

♣ Every day over 4,000 people, mostly women and children, die as a result of long exposure to air pollution resulting from the smoke from the inefficient conversion and use of biomass fuels,

♣ In almost all the cases in sub-Saharan Africa, the demand for biomass fuels far exceeds sustainable supply. Among the resulting environmental impact are varying degrees of deforestation, general land degradation, soil erosion, loss of soil fertility and desertification,

♣ Dwindling biomass resources has led to additional workload for women and children, as they spend longer hours sourcing cooking energy, at the expense of other productive endeavors, including education of the latter.

However, in spite of these and other negative externalities, and in the light of the reality of their economic situations, the use of biomass fuels by the poor has been identified with a few advantages that cannot be ignored. Among these are that:

❖ To a large extent, and in the long term, biomass fuel can be a renewable source of energy,

❖ Biomass fuels are available in some form (farm residue, dung, fuel wood) virtually ubiquitously, and can be converted to energy (burned) without further processing,

❖ Biomass fuels are affordable to the poor because they are cheaper than such other alternative fuels as electricity or gas or paraffin,

❖ Technologies and techniques for the sustainable production and use of biomass energy are either available or can be readily made available.

Against this background, the International Energy Agency has cautioned that without strong new policies to expand access to cleaner fuels and technologies, one-third of the world's population are projected to still be relying on biomass fuels in the next few decades. Furthermore, the Agency asserts that there is evidence that, in areas where local prices have adjusted to recent high international energy prices, the shift to modern, clean and more efficient use of fuel for cooking has seen very sharp reversals (IEA, 2006). A number of matters arise from this scenario. They include:

1. The overwhelming dependence of the poor on the use of biomass fuels for cooking would not be an issue of grave concern, if the base resources are harvested in a sustainable manner, on the one hand, and if the harvested biomass is converted to cooking energy in an efficient and environmentally friendly manner.

2. Following from the above, is the need to explore ways and means of improving the way that biomass fuels are supplied and subsequently used for cooking. This is so because it does not appear feasible to switch from traditional sources of cooking energy, at least in the short term.

3. The quality of any policy on the regulation and control of the use of biomass fuels will be greatly influenced by the quality of the information on which it is based. In other words, detailed and accurate statistics on energy supply and consumption are essential for proper

policy and market analysis. It is in this light that the paucity of data and statistics constitutes a major challenge.

Policy Challenges

Although dependence on biomass fuels as the major source of cooking energy by the poor may be considered rational, it is clear that this preference cannot be sustained, due to some of their environmental impact, including those highlighted in this paper. However, changing the cooking habits of the poor has never been, and can never be, an easy task. Users of biomass fuels must be convinced that there are better methods than the traditional ways. As we approach the eve of the target (2015) set by the United Nations Millennium Project, and considering the very close links that cooking energy has with each of the Millennium Development Goals (MDGs), on the one hand, and the need to address the macro and micro environmental challenges associated with the use of biomass fuels as cooking energy, on the other, it becomes extremely urgent that we look closely at some of the policy implications of the issues discussed in this paper. As a minimum it is strongly recommended that a responsive policy on the cooking energy needs of the poor in sub-Saharan Africa must include the following, sometimes interrelated and even over lapping items: an aggressive environmental education; the identification of a pragmatic alternative; capacity building for its production and commercialization; the question of affordability and the data need.

Environmental Education: Sensitization and Awareness

If, as it has been confirmed by various investigations, switching from traditional biomass fuels to modern sources cannot be feasible in the short run for many households, then the success of any proposed option aimed at that switch will depend strongly on its social acceptability. Social acceptability, in turn, will depend on how much prospective users of the improved methods know about the disadvantages of the old and the advantages of the new forms.

Massive and aggressive environmental education must, therefore, be a strong component of any sustainable long term policy and strategy for redressing the stresses that result from biomass energy consumption. Awareness about the environmental implications of dependence on traditional sources of cooking energy can be raised through public and private agencies, grassroots institutions, through changes in school curricula and by creative use of the local media. Every perceived advantage must be explored to sensitize, mobilize and educate the biomass-energy-dependent poor on why the change of the source of cooking energy must be considered, and an alternative or alternatives embraced.

Considering that the vast majority of the target populations reside in rural areas, and considering the enormity of the power they wield, and the control they have over their subjects in Nigeria, local institutions built around traditional rulers and royal fathers, such as the emirs, obas, obis, igwes and chiefs, etc, constitute very potent platforms for grassroots mass sensitization and mobilization for the social acceptability of new energy schemes. Similarly, considering the great influence and followership of clerics, religious institutions such as churches and mosques can also be veritable outlets for the dissemination of environmental information.

Environmental education should be made a mandatory subject in the curricula of primary and secondary education. Considering also that the majority of users of biomass fuels are illiterate, environmental education must be reduced or translated into the local languages of the communities for maximum dissemination and assimilation. Policy makers and designers of environmental education must ensure that information to be disseminated must be rendered in simple, easily understandable language, completely devoid of all confusing technicalities.

In addition, professional groups and non-governmental organizations (NGOs) should be encouraged and indeed sponsored and supported by various levels of government and donor agencies to organize environmental awareness programmes for biomass fuel users. Because of its national significance and importance, even political parties should be compelled to include environmental education in their manifestoes.

However, environmental education must recognize that although the rural poor are most directly implicated in the use of biomass fuels, the impact often goes far beyond their immediate communities. For instance, urban consumers of fuel wood are scarcely aware of the consequences (even if only indirect) of increased demand for soil fertility and eventually food supply.

Attributes of the Alternative

The second policy issue is the content of the sensitization programme. It is presumed that the introductory part of the environmental education programme would address the negative externalities of the present form of biomass energy consumption, which become the basis for advocating a switch to alternative forms. In other words, present users of the current forms of cooking energy must understand those attributes of the alternative forms that make them better and preferred. Essentially, these attributes should be those that neutralize the perceived negative impact of traditional biomass energy. Consequently, the policy should ensure that the attributes of the improved alternative(s) include, but by no means limited to the following:

> A remarkable saving of the quantity of biomass energy used for the same amount of cooking. This saving would be coming from the increased efficiency of conversion, thereby achieving MDG 7,

> A remarkable reduction in indoor pollution, arising principally from a complete combustion process of the conversion of biomass, thereby reducing suspended particulate matter, and, in turn, reducing the risk of respiratory diseases and of eye infections, and thereby reducing child mortality, improving maternal health and combating diseases, ultimately achieving MDGs 4, 5, 6 and 7,

> A remarkable improvement in the design and ventilation of kitchens, such that even with the existing form of biomass use, the particulate levels in the homes using fuel wood could be remarkably reduced. This is claimed to have reduced particulates to levels lower than in homes using LPG (IEA, 2006),

> An appreciable reduction in health risks (such as burns and cough, among others) and time-loss by women and children, which could then be used for some productive, income earning activities, thereby reducing child mortality, improving maternal health and combating disease, and ultimately achieving MDGs 1, 4, 5 and 6,

> An unambiguous reduction in emission of carbon monoxide and other greenhouse gases into the atmosphere, thereby achieving MDG 7,

➢ The employment opportunities that would arise from the jobs and small scale businesses that would be created through the production and commercialization of improved forms of biomass energy use, thereby achieving MDG 1,

➢ The overall economic empowerment that would arise from the improved incomes that would be earned from the production and commercialization of improved forms of biomass energy use, thereby eradicating extreme poverty and hunger, and ultimately achieving MDG 1,

➢ The reduced use of dung and agricultural residue will, with time, improve soil fertility and contribute to a reduction of land degradation. In the same vein, a reduction of the demand for, and consumption of, wood fuel will in turn reduce pressure on forest resources. This will in turn result in savings of resources that would have been used for afforestation. Ultimately, both will ensure environmental sustainability, thereby achieving MDG 7.

The Alternative

There are a wide range of technologies for modern cooking. They range from artisanal or factory-made clay and metal stoves to solar cookers. Others include heat retainers as well as stoves using modern biofuels, such as plant oil, ethanol or biogas (IEA, 2006; 6). Considering that we are concerned with the poor in sub-Saharan Africa, most of these technologies are only relevant intellectually but not practically, because of affordability. Caution must be exercised in recommending an alternative or alternatives; considering that per capita income in the region is not expected to rise remarkably, if all, in the very near future, and also that cooking energy consumption is very sensitive to price changes.

It is against this background that the ***Rocket Stove*** is recommended as a replacement for the traditional and ubiquitous three-stove open fire. The Rocket Stove is designed by GTZ in conjunction with the Aprovecho Institute. It was invented by Dr. Larry Winiarski, the present Director of the Institute, in 1982 (http://www.ashdenawardsorg/winners/trees). It won the Ashden Award for sustainable energy in 2005, in the "Health and Wealth" category (http://www.stovesource.com/mambo/index.php?option=comcontent&task=category&sectioned=5&id=20&Itemid=51). It also won the Special Africa Award at the Ashden Award in 2006 for their work with rocket stoves for institutional cooking in Lesotho, Malawi, Uganda, Mozambique, Tanzania and Zambia (http://www.ashdenawards.org/winners/aprovecho).

There is a variety of the stove. Some are fabricated from metal, while others are made of clay. Some (like those made of metal) are mobile, while others (especially those made of clay) are fixed and stationary. The underlying principle is, however, the same. The Rocket Stove has a unique fire chamber, where virtually all the gases that are produced when the fire is lit are completely burnt up within the chamber and the heat is transferred efficiently and effectively to the pot.

The Rocket Stove combines air-intake with the fuel feed slot. It is made up of four components as follows:

♣ ***The Fuel Magazine***: into which the raw, unburned fuel (e.g. wood or dried grass) is loaded, fed or placed, and from where it feeds into the combustion chamber,

♣ ***The Combustion Chamber***: located at the end of the fuel magazine, where the raw fuel (wood, etc) is burned,

♣ *The Chimney*: A very vertical channel above the combustion chamber, to provide the updraft of air needed to maintain the fire,

♣ *The Heat Exchanger*: To transfer the heat to where it is needed, that is, the cooking pot (en. wikipedia.org/wiki/Rocket_stove).

If the wood that is fed into the stove is thoroughly dried, the Rocket Stove ensures almost complete combustion, thereby reducing the emission of suspended particulate matter and reducing all the attendant health risks. Since none of the energy converted from the wood escapes, the stove saves energy, thereby reducing waste and cooking time, as well as reducing the wood needed to generate the required cooking energy. The reduction in the wood energy fed into the stove in turn, reduces the environmental pressures on vegetation and soils; reduction of greenhouse gasses, among the other attributes listed earlier. The Rocket Stove has been successfully deployed in more than 100 projects in over 60 countries, including those in East Africa, especially Uganda, Ethiopia and particularly in Rwanda, where it was used in cooking in large refugee camps (www.cd3wd.com/cd3wd_40/ JF/425/20-228.pdf; www.aprovecho.org/).

There are clear and specific multi-level benefits of the improved Rocket Stove, ranging from the individual family, the local community, the nation to the global community. They have been summarized to include:

❖ *Potential benefits to the family:*
  ➢ Less time spent gathering wood or less money spent on fuel,
  ➢ Less smoke in the kitchen,
  ➢ Lessening of respiratory problems associated with smoke inhalation,
  ➢ Less manure used as fuel,
  ➢ Releasing more fertilizer for agriculture,
  ➢ Little initial cost, compared to most other kinds of cookers,
  ➢ Improved hygiene with models that raise cooking off the floor,
  ➢ Safety: fewer burns from open flames;
  ➢ Less chance of children falling into the fire or boiling pots;
  ➢ If pots are securely set into the stove, less chance of children pulling them down on themselves,
  ➢ Cooking convenience: stoves can be made to any height and can have work space on the surface,
  ➢ The fire requires less attention, as stoves with damper control can be easier to tend.

❖ *Potential benefits to the local community:*
  ➢ Stove building creates new jobs,
  ➢ Potential for 100 per cent local content (materials),
  ➢ Potential for local innovations,
  ➢ Money and time saved can be invested elsewhere in the community.

❖ *Potential benefits to the nation:*
  ➢ Lowered rate of deforestation improves climate, wood supply and hydrology,
  ➢ Decreases soil erosion,
  ➢ Potential for reducing dependence on imported fuel,

> ➤ Potential for short-range solution to deforestation, while long-range reforestation programmes get underway,
> ➤ Cost of providing Rocket Stove is low, compared with other means of fighting deforestation.

❖ *Potential benefits to the global community:*
> ➤ Stoves can slow down the rate of climate change, deforestation and desertification,
> ➤ They allow time for reforestation projects to gain a foothold and help to change the balance toward extending forested areas once again (Aprovecho Institute, 1984; 12).

The appeal of the Rocket Stove is, therefore, not debatable. It has been tried. It has been tested. It has been confirmed to work. It provides the short run solution to the impending 'other energy crisis in developing countries': the shortage of wood fuel.

Capacity Building

Given the disproportionately large population that depend on biomass energy for cooking, a successful sensitization programme and wide-spread social acceptability would mean a huge demand for the proposed alternative, improved and more efficient form of converting biomass fuels into cooking energy. There would, therefore, the need to build the needed capacity to cope with this expected huge demand.

A responsive policy would, therefore, require governments at all levels to make, or improve provision for the training of locals for the purpose of preparing the people to receive the new concept, as well as developing the needed skills and expertise. People need to be trained to educate prospective users about the advantages of the new alternatives. People also would be needed to be trained on how to produce and commercialize them.

The capacity of the end user of the new alternative (especially the women and children who do most of the cooking) also needs to be enhanced. The objective here is to ensure a more efficient use of biomass energy. For instance, letting biomass energy users know that drying fuel wood thoroughly before feeding it into the fuel magazine of the Rocket Stove can reduce smoke levels, is an aspect of capacity building. Similarly, letting people know that placing the lid on the pot during cooking can reduce cooking time, thereby reducing the amount of energy needed to cook a particular food, as well as the amount of smoke that would be released during cooking, are aspects of capacity building. Furthermore, letting people know that increasing the number of window openings in the kitchen, or providing gaps between the roof and wall, as well as moving the cooking stove, if it is the mobile type, away from the living area of the house can enhance environmental quality and reduce health risks, are all aspects of capacity building.

Economic Affordability

On Friday, January 28, 2011, a Nigerian daily national newspaper carried a picture of a man processing traditional biomass (wood fuel) for sale. The caption on the picture was: "Firewood business booms as scarcity of kerosene persists in Abuja" (Punch, 28.01.11; 13). Similarly, IEA (2006; 441) reported of Brazil, that "biomass consumption per capita has stopped declining and even started to increase, as many poor households switch back to fuel wood, in the face of higher LPG prices". This is to show that the issue of affordability of cooking energy is a global one. The bottom line of any policy aimed at persuading the poor to switch from the traditional forms of biomass consumption to a cleaner and more efficient form is to ensure that the new alternative is affordable.

One of the anticipated barriers to the acceptance and subsequent penetration of improved biomass energy use methods is the high cost of fabricating and or buying the basic unit. If per capita incomes are not expected to rise appreciably, and the poor do not normally have adequate access to facilities to help upgrade their economic status, then governments must set up structures to assist them. Compared to the international response to hunger, HIV/AIDS, dirty water, poor sanitation and malaria, energy use for cooking, especially from the point of view of the poor, has received very woeful funding and political backing. Yet considering the direct and indirect collateral environmental damages that traditional biomass energy consumption has caused and will continue to cause, the urgent extension of funding assistance cannot be overemphasized and is strongly advocated.

One approach is to encourage the development of micro financing. Microfinance institutions allow households and villages to mobilize and access the capital needed to make small energy investments by the managers of cooking in the home; the women. IEA (2006; 443) has established that worldwide, four out of every five micro-borrowers are women. Because the amounts involved are small, no collaterals are needed and the repayment terms are liberal, micro financing is attractive even in the rural areas, where biomass energy consumption is highest. Governments at all levels, non-governmental organizations (NGOs) and international donor agencies should be encouraged to partner with households and village communities to address the perennial cooking energy needs of the rural poor.

Tooling for Policy

The challenges facing the poor concerning meeting his cooking energy consumption vary from country to country and even from community to community within the same country. There is need, therefore, for policies to be locally customized to reflect such differences and peculiarities. One of the major challenges in this task is the lack of quality data in terms of geographic (spatial) comprehensiveness and temporal currency. There is, therefore, the need to develop robust national, regional and local data bases, which should include as the minimum, information on the following: population to be served, household sizes, rural and urban populations, cooking frequencies of certain common local meals, per capita incomes, the sources of the potential fuels locally available, health statistics, hospital records of certain diseases, vegetation and soil conditions of sources of local fuels, the types of stoves and other cooking devices, available infrastructure and prospective producers, cost and market analyses, estimates of the ability and willingness to pay for improved facilities as a function of incomes. In order to make such data relevant for development planning, they must reflect local differences. They, therefore, must be geo-referenced and converted to geographic information systems (GIS) formats. This will ensure their comprehensiveness and ease of retrieval. The policy must also include provision for the regular update of energy information.

5   Concluding Remarks

In this paper, we have established that since the late 1960s, sub-Saharan African leaders have recognized the importance of the environment and its resources to the development of their economies. Consequently, they have signed and adopted a number of conventions to protect and conserve their environments. We have also established that at the level of individual nations, myriads of environmental instruments have also been adopted to protect the environment so that it can stimulate, enhance and continue to sustain socio-economic development. However, in spite of these continental and national instruments, environmental have not only persisted, but even deteriorated in certain cases.

Although the direction and strength of the factors operating within the nexus do not appear to be very clear, the fact that there are very strong links between economic status and environmental quality is no longer in doubt. Within the nexus, the strong links between poverty and the environment have also been well established. At a lower level of resolution, the strong links between cooking energy consumption and the environment are also no longer controvertible. The very core significance of cooking energy to the prosecution of the United Nations Millennium Project has been demonstrated by the close links between the former and each of the goals set by the latter. However, as a result of widespread poverty, it has been projected that household cooking energy preferences will likely experience inertia in the next several years. In other words, since per capita incomes in sub-Saharan Africa are not expected to witness a quantum leap in the near future, it would be unrealistic to expect or recommend a revolutionary switch from traditional forms of energy consumption to the cleanest but most expensive forms. While we prepare to embrace new technologies in cooking energy, even marginally improved facilities will suffice for now.

However, any recommended intermediate technology must be such that is capable of dealing with most of the perceived challenges of the traditional forms and methods. Consequently, it should be able to reduce the quantity of biomass energy used through more efficient conversion; reduce indoor pollution as a result of a more complete combustion process; improve the design and ventilation of kitchens; reduce health risks and time-loss by women and children; reduce suspended particulate matter, on the one hand, and the emission of carbon monoxide, and other greenhouse gases into the atmosphere, on the other; create employment opportunities; result in economic empowerment; reduce land degradation by leaving dung and farm residue on the field, and make optimal contribution to the achievement of the United Nations Millennial Development Goals (MDGs).

The product that appears to combine all these attributes is the Rocket Stove, designed under the auspices of GTZ. It has been widely accepted for use in many developing nations, including some East and Southern African countries, with very great results. For instance, within two years of its introduction and adoption, the rocket stove has reached over 200,000 households, generated over 290 jobs, which in turn, generated incomes of about 261,000 Euros per annum, in Uganda. Furthermore, with the use of improved stove over 200,000 tons of wood are saved very year, which is the equivalent of spending almost eight million Euros in afforestation (GTZ, 2007). It, therefore, holds great prospects for large countries like Nigeria that are under the pressure of a large and fast growing population. However, in order to popularize it in other sub-Saharan African countries, an aggressive environmental education is needed. Local capacity also must be built to both produce and commercialize it. Finally, a strong micro finance structure must be established to support the scheme and assist those who may have initial difficulty in embracing the new, improved biomass cooking energy technology.

REFERENCES

Adeoti, O., Idowu, D. O. O. and Falegan, T. (2003), "Could Fuel Wood Contribute to Household Poverty in Nigeria?", *Biomass and Energy,* Vol. 21, pp. 205-210.

Aggrey, N., Wambugu, S., Karugia, J and Wanga, E. (2010), "An Investigation of the Poverty-Environment Degradation Nexus: A Case Study of Katonga Basin in Uganda" *Research Journal of Environmental and Earth Sciences*, Vol. 2, No. 3, pp. 82-88.

Amechi, E. P. (2009), "Poverty, Socio-Political Factors and Degradation of the Environment in sub-Saharan Africa: The Need for a Holistic Approach to the Protection of the Environment and Realization of the Right to Environment", *Law, Environment and Development (LEAD) Journal*, Vol. 5, No. 2, pp. 109-129.

Aprovecho Institute (1984), *Fuel-Saving Cook Stoves*, A German Appropriate Technology Exchange (GATE) Publication, 129 pp.

Aprovecho Research Centre (ARC) (2006), *Rocket Stoves for Sustainable Cooking*, South Africa: Ashden Awards.

www.aprovecho.org/ Aprovecho Research Centre (ARC) (2009), *Applying Appropriate Technology to Improve Lives and the Environment.*

www.unep.org/Nairobi Convention.docs/English Nairobi ConventionText.pdf.

http://www.opeworg/chemica-weapon-convention/related-international agreements/toxic-chemica-weapons-and-the-environment/marine-and-coastal-environment-west-and-centra-africa/

Arnold, M. Kohlin, G. Persson, R. and Shepherd, G. (2003), "Fuel Wood Revisited: What has Changes in the last Decade" Centre for International Forest Research (CIFR), *Occasional Paper, No 39.*

Bucknall, J., Kraus, C. and Pillai, P. (2000), *Poverty and the Environment*, The World Bank, (April).

Chavin, H. (1981), *Une Ville Africaine en Crise d'Energie: Ouagadougou*, Edition Unasylva.

Chukwu, G. U. (2008), "Poverty-Driven Causes and Effects of Environmental Degradation in Nigeria", *The Pacific Journal of Science and Technology*, Vol. 9, No. 2, pp. 599-602.

Comim, F. (2008), *Poverty and Environment Indicators*, UNDP/UNEP: Capability and Sustainability Centre.

Department for International Development (DFID), (2000), *Achieving Sustainability: Poverty Elimination and the Environment: Strategies for Achieving the International Development Targets*, DFID Plans.

Detwyler, T.R. and M.G. Marcus, (1972), Urbanization and Environment in Perspective" in T.R. Detwyler and M.G. Marcus (eds.), *Urbanization and Environment*, Bermont, California: Duxbury Press, pp. 3-26.

Duetsche Gesellschaft fur Technische Zusammenarbeit (GTZ), (1979), *Helping People in Poor Countries Develop Fuel-Saving Cook Stoves*, Germany: A German Appropriate Technology Exchange (GATE) Publication. 148 pp.

Duetsche Gesellschaft fur Technische Zusammenarbeit (GTZ), (2007), *Cooking Energy: Why it really Matters if we are to Halve Poverty by 2015*, (23 pp.).

Duraiappah, A. (1996), *Poverty and Environmental Degradation: A Literature Review and Analysis*, London: International Institute for Environment and Development.

Ezzati, M. and Kammen, D. (2001), "Indoor Air Pollution from Biomass Combustion as a Risk Factor for Acute Respiratory Infections in Kenya: An Exposure-Response Study", *Lancet,* Vol. 358(9281), pp. 619-624.

Fawehinmi, A. S. and Oyerinde, O. O. (2002), "Household Energy in Nigeria: The Challenge of Pricing and Poverty in Fuel Switching", *Journal of Energy and Development*, Vol. 27, No. 2, pp. 277-284.

http://www.stovesource.com/mambo/index.php?option=comcontent&task=category&sectionid=5&id=20&Itemid=51.

http:/www.ashdenawards.org/winners/trees, *Just stove wins award.*

Federal Republic of Nigeria (2004/2007), *Federal Environmental Protection Agency (FEPA) Act, Cap F10, FLN, 2004*, now superseded by *National Environmental Standard and Regulations Enforcement Agency (Establishment) Act, No. 25 of 2007.*

Federal Republic of Nigeria (2004/2007), *Federal Environmental Protection Agency (FEPA) Act, Cap F10, FLN, 2004*, now superseded by *National Environmental Standard and Regulations Enforcement Agency (Establishment) Act, No. 25 of 2007.*

Food and Agricultural Organization (FAO) (2003), "Forestry Outlook Study for Africa: Regional Report-Opportunities and Challenges Towards 2020 Food and Agricultural Organization of the United Nations", *Forestry Paper 141.*

Gundimeda, H. and Kohlin, G. (2003), *Fuel Demand Elasticities for Energy and Environmental Studies: Indian Sample Survey Evidence*. Environmental Economics Unit, Department of Economics, Goteborg University, Sweden.

Holden, S., Bekele, S. and Pender, J. (2004), "Non-farm Income, Household Welfare and Sustainable Land Management in less-favoured Areas in Ethiopia", *Food Policy*, Vol. 29, No. 4, pp. 369-392.

International Energy Agency (IEA) (2006), "Energy for Cooking in Developing Countries", in OECD/ IEA (2006), *World Energy Outlook 2006: Focus on Key Topics*, (Chapter 15), pp. 419-445.

Johnson, S. P. (ed.) (1993), *The Earth Summit: The United Nations Conference on Environment and Development, (UNCED)*. London: Graham and Trotman.

Joseph, K. and Sumption, S. (1979), *Equity,* London: J. Murray Publications.

Lowe, M. S. and Bowlby, S. R. (1992), "Population and Environment" in Mannion, A. M. and Bowlby, S. R. (eds.) *Environmental Issues in the 1990s*. New York: John Wiley and Sons.

Malthus, T. R. (1798), *An Essay on the Principle of Population*, Penguin Classics (Reprinted 1985).

Molb, M., Johl, A., Wagner, J. M., Popovic, N., Lador, Y., Hoenniger, J., Seybert, E. and Walters, M. (eds.) (2005), *Earth Justice, Environmental Rights Reports: Human Rights and the Environment*. Oakland: Earth Justice (Because the Earth needs a good Lawyer) (Appendix), pp. 86-108).

Nolan, B. and Whelan, C. T. (1996), *Resources Deprivation and Poverty*, Oxford: The Clarendon Press.

www.policy.co.uk/key concepts/samples/lister-chapter.pdf.

Omuta, G.E.D. (1988), "The Quality of Urban Life and the Perception of Livability: A Case Study of Neighbourhoods in Benin City, Nigeria", *Social Indicators Research*, 20, pp. 417-440.

Omuta, G.E.D. (2004), *The Good, the Bad and the Ugly: Man-Nature Interfaces*, The Seventy-fourth Inaugural Lecture, University of Benin, Benin City.

Organization of African Unity (OAU), (1968), *African Convention on the Conservation of Natural and National Resources*, Algiers: September 15, (Doc. No. CAB/LEG 24.1).

Osei-Hwedie, K. (1995), "Poverty and Environment: Dimensions of Sustainable Development Policy", *PULA: Botswana Journal of African Studies*, Vol. 9, No. 2, pp. 1-10.

www.worldenergyoutlook.org/docs/weo2006/cooking.pdf.

Ostrom, E., Burger, J., Field, C. B., Norgerd, R. B. and Policansky, D. (1999), "Sustainability: Revisiting the Commons; Local Lessons, Global Challenges", *Science*, Vol. 284, pp. 278-282.

Ouedraogo, B. (2005), "Household Energy Preferences for Cooking in Urban Ouagadougou, Burkina Faso", Elsevier: *Article in Press in Energy Policy.*

www.bm.kbase.com/articles/LPGsdarticles(15).pdf.

Putnam, R. D., Leonardi, R. and Nanetti, R. Y. (1993), *Making Democracy Work: Civic Traditions in Modern Italy*, Princeton: Princeton University Press.

Ravallion, M. (1996), "Issues in Measuring and Modeling Poverty", *The Economic Journal*, Vol. 106 (September), pp. 1328-1343.

Republic of South Africa (1998*), National Environmental Management Act (NEMA), No. 107*.

Republic of Kenya (1999), *Environmental Management and Coordination Act (EMCA), No. 8*

Ringen, S. (1987), *The Possibility of Politics*, Oxford: The Clarendon Press.

Sen, A. K. (1981), *Poverty and Famines: An Essay on Entitlement and Deprivation*. New Delhi: Oxford University Press.

Scott, L. (2006), *Chronic Poverty and the Environment: A Vulnerability Perspective*. London: Overseas Development Institute (ODI), (August).

Smith, S. D., Huxman, T. E., Ziter, S. F., Charlet, T. N., Housman, D. C., Coleman, J. S., Fenstermaker, L. K., Seemann, J. R. and Nouak, R. S. (2000), "Elevated CO_2 increases Productivity and Invasive Species Success in an Arid Ecosystem", *Nature*, Vol. 408 pp. 79-83.

Sow, H. (1990), *Les bois-Energie au Sahel*, Paris: ACCT-ATC-KARTHALA.

Somaratne, W. G. (2005), "Globalization, Poverty Alleviation and the Environmental Governance: The Strategic Policy and Operational Options for South Asia", *SAARC Journal of Human Resource Development*, pp. 55-71.

The Concise Oxford Dictionary (1995).

Tolba, M. K. El-Kholy, O. A., Holdgate, M. W. MacMicheal, D. F. and Munn, R. E. (1992), *The World Environment: 1972-1992; Two Decades of Challenge*, London: Chapman and Hall.

Townsend, P. (1979) *Poverty in the UK: A Survey of Household Resources and Standards of Living*. Berkeley and Los Angeles: University of California Press.

www.tinygreenbubble.com/livegreen/.../546-clean-energy-definition.

dictionary.reference.com/browse/clean + energy.

www.wisegeek.com/what-is-clean-energy.htm.

www.Twitter.com/TinyGreenBubble.

File///C:Users/Owners/Desktop/clean_energy_definition.html.

United Nations Fund for Population Activities (UNPFA), (1991), *Population, Resources and Environment: The Critical Challenges*. London: Banson Productions.

Wason, D. and Hall, D. (2004), "Poverty in Lesotho: 1993-2002; An Overview of Household Economic Status and Government Policy", *Critical Poverty Research Centre Working Paper, No. 40*, London: Overseas Development Institute (ODI).

Wilkinson, F.F and M. Wyman (eds.), 1986, Environmental Challenge: Learning for Tomorrow's World, London: Althouse Press.

World Energy Outlook (WEO) (2006), "Energy for Cooking in Developing Countries", in *WEO 2006: Focus on Key Topics*, (Chapter 15), pp. 419-445.

www.gtz.de/de/dokumente/en-cooking-energy-2007.pdf

www.cd3wd.com/cd3wd_40/JF/425/20-448.pdf.

www.gate-international.org/documents/publications/webdocs/.../g14fue.pdf.

en.wikipedia.org/wiki/Rockect_stove.

http://www.ashdenawards.org/winners/aprovecho."